The Technology of Instrument Transformers

Ruthard Minkner · Joachim Schmid

The Technology of Instrument Transformers

Current and Voltage Measurement and Insulation Systems

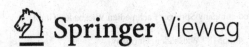

Springer Vieweg

Ruthard Minkner
Pfeffingen, Switzerland

Joachim Schmid
Müllheim, Germany

ISBN 978-3-658-34865-6   ISBN 978-3-658-34863-2   (eBook)
https://doi.org/10.1007/978-3-658-34863-2

Responsible Editor: Reinhard Dapper
This Springer Vieweg imprint is published by the registered company Springer Fachmedien Wiesbaden GmbH part of Springer Nature.
The registered company address is: Abraham-Lincoln-Str. 46, 65189 Wiesbaden, Germany

Foreword

During the work on the monograph "Ferroresonance Oscillations in Substations with inductive voltage transformers, in medium and high voltage systems" we came across the different design principles of instrument transformers, which are not described in the literature. In the German technical literature concerning instrument transformers, the book by R. Bauer [1], published in 1953, and a collection of essays, published by I. Grambow [2] in 2003, should be mentioned. In addition, individual aspects of instrument transformers are dealt with in articles in technical journals. Furthermore, the use of electric fields [3] in the dimensioning of instrument transformers is missing in the current literature. The use of this tool was only made possible by availability of high-performance computers. In the present textbook electric fields play an essential role and enable designs of instrument transformers which were not possible 10 years ago.

The two authors Dr.-Ing. Ruthard Minkner and Dr.-Ing- Joachim Schmid are active in the field of high voltage measurement technology.

Dr.-Ing. Ruthard Minkner was director of the company Emil Haefely AG in Basel. He has industrial experience in high voltage systems and in stability problems of non-linear systems. He taught high voltage engineering and control engineering at the Bern University of Applied Sciences in Burgdorf and was a visiting professor at the State University of Washington (USA) in Pullman. He led various IEC working groups and received the "IEC 1906 Award" three times.

Dr.-Ing. Joachim Schmid was head of development for instrument transformers at the company Trench Switzerland AG in Basel. He is chairman of the Swiss Electrotechnical Committee CES TK 38 and heads various IEC working groups. He was also presented with the "IEC 1906 Award" for his commitment in the creation of the standards. He supervised several diploma theses at the Bern University of Applied Sciences in Burgdorf in the field of low-power instrument transformers.

The sections on new measurement technologies also cover the optical current transformer and the RC divider for voltage measurement. For voltage measurement, the compensated R-divider and the RC-divider seem to become accepted as the most modern measuring device.

Many open questions were clarified with the help of diploma theses written by students at the Bern University of Applied Sciences in Burgdorf. J. Schmid, R. Minkner and V. Karius were the contact persons for the diploma students. A copy of the diploma theses can be found at the Bern University of Applied Sciences. The results taken from them are the intellectual property of the diploma students and supervising experts.

The authors would like to thank the Bern University of Applied Sciences in Burgdorf for the opportunity to solve technically unexplained phenomena in the context of semester and diploma theses.

Ruthard Minkner had many discussions about existing difficulties in high voltage measurement technology during the lecture on high voltage measurement technology at the State University of Washington in Pullman. We would like to thank PhD Edmund Schweizer, President of the Relays Company SEL for his interest in high voltage measurement.

Springer Verlag is thanked for the willingness to publish this textbook.

2020   Ruthard Minkner
                                                              Joachim Schmid

Foreword to the English version

During the work on this book, it was always the wish of Ruthard Minkner to also publish an English version. Unfortunately, he passed away just after the German book was published.

The translation of the book into English language was done by Joachim Schmid. Some mistakes in the German version had been corrected in this English version.

Further to the acknowledgements in the original foreword I would like to thank Ruthard's wife Dorothea Minkner for her tremendous help during the work on this book.

2021   Joachim Schmid

Introduction

The authors endeavour to make the content as comprehensible as possible to the reader by using practical examples and detailed illustrations.

Table 1 shows an overview of the contents of the book.

Table 1 Overview of the topics covered

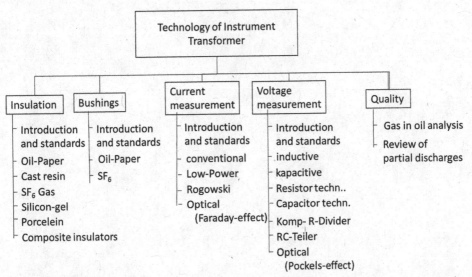

In this textbook, designs and calculations are recorded, which have been brought into regulations, training courses and lectures by the authors.

A special chapter was dedicated to the technology of resistors used in compensated R-dividers and RC-dividers. The use of resistors combined with capacitors in voltage transformers is only at the beginning.

The inspection of instrument transformers during operation is today limited to partial discharge measurement, gas in oil analysis, leakage etc. These points are dealt with in the chapter Quality Control.

Contents

Insulation for High Voltage Equipment

1

1.1 Introduction and standards

The following chapter describes the electrical insulation of instrument transformers. Both the internal insulation of the instrument transformer with different materials and the external insulation with porcelain or composite insulators are covered.

Liquid, gaseous, and solid insulating materials are used for internal insulation. An essential aspect for the evaluation of the materials is the test arrangement used, which differs from the IEC arrangement (sphere—sphere) [4]. A test arrangement as shown in Fig. 1.1a was used for the oil-paper insulations described in the following chapters. Figure 1.2b shows an advanced electrode with a Rogowski profile. The high voltage electrode cast in epoxy resin prevents an increase in field strength at the edge. With these electrode arrangements, it is possible to measure the dielectric strength or the breakdown voltage for the insulation systems used. As a liquid impregnating fluid, the insulating oil Shell Diala D hydrofinish is used as an example. The experience of the authors is based on investigations with this oil, which is no longer produced today.

The described test device can be used for liquid, solid and gaseous insulating materials. For solid insulating materials, layered oil-paper insulations, layered film insulations and solid insulating materials such as cast resin are used.

For the insulation of instrument transformers, the two international standards IEC 60071–1 [5] and IEC 60505 [6] are essential.

© Springer Fachmedien Wiesbaden GmbH, part of Springer Nature 2022
R. Minkner and J. Schmid, *The Technology of Instrument Transformers*,
https://doi.org/10.1007/978-3-658-34863-2_1

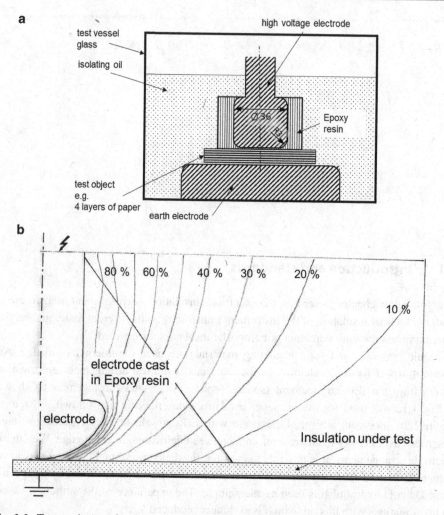

Fig. 1.1 Test equipment for determining the dielectric strength **a** Arrangement for determining the dielectric strength of insulating materials, e.g. oil-impregnated paper insulation. **b** Electrode arrangement with cast-in Rogowski profile for determining the dielectric strength of insulating materials

IEC 60071–1 describes the insulation coordination of insulation systems in high-voltage equipment. As a function of the highest permissible system voltage U_m, values for overvoltages are defined which the insulation must withstand. For devices with U_m less than 300 kV the AC withstand voltage and the lightning impulse withstand voltage (BIL) are defined, for devices with U_m greater than or equal to 300 kV the lightning impulse withstand voltage and switching impulse withstand voltage (SIL) are defined. IEC 60071–2 [7] provides application guidelines for insulation coordination.

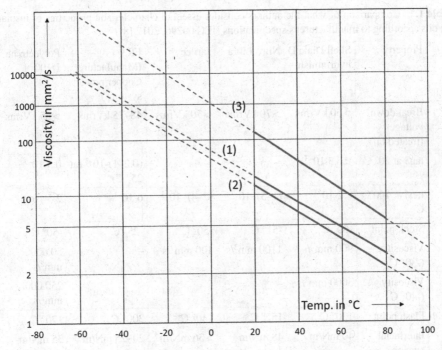

Fig. 1.2 Viscosity of Shell Diala D hydrofinish insulating oils (1), Polyectren D12 (2) and Polyectren D100 (3)

IEC 60505 describes the basis for determining the lifetime of an electrical insulation system under electrical, thermal, mechanical, and environmental influences. The mechanisms of ageing are explained and test methods for lifetime evaluation are described.

1.2 Insulating oils for instrument transformers

1.2.1 Properties of different insulating oils

The oils listed in Table 1.1, "Shell Diala D hydrophinish and "Nytro Libra" from Nynas are refined mineral oils without additives. Most of the authors' experience is available for the mineral oil Shell Diala D hydrofinish, which has been used for over 20 years. This oil is no longer produced today. The disadvantages are fluctuations in the base oils.

The synthetic oils Jarilec and Polyectren D100 have the advantage of a constant composition. The disadvantages compared to Shell Diala D hydrofinish are the more than 10 °C higher pour point of Jarilec and Polyctren D 100.

The IEC standard IEC 60296 [8] specifies some important requirements that apply specially to instrument transformers that have practically no intrinsic temperature rise:

Table 1.1 Comparison of what the authors consider essential characteristic properties of insulating oils according to manufacturer's specifications (IEC 60296, 2012 [8])

No	Property	Shell Diala D hydrofinish	Nitro Libra	Jarilec	FR3 (Manufacturer: Cooper power systems)	Polyectrene D100
1	Breakdown voltage (treated oil)	>60 kVrms	>70 kVrms	>70 kVrms	48–75 kVrms	>65 kVrms
2	tanδ at 20° C	$<0.3 \cdot 10^{-3}$	–	–	$(0.2–1) \cdot 10^{-3}$ (at 25° C)	$0.3 \cdot 10^{-3}$
3	tanδ at 100° C	$<2 \cdot 10^{-3}$	$(1–5) \cdot 10^{-3}$	$(2–4) \cdot 10^{-3}$	$6 \cdot 10^{-3}$	$2.5 \cdot 10^{-3}$
4	Stock point	-60° C	-51° C	-50° C	-25 °C	-50° C
5	Viscosity (-30° C)	900 mm²/s	1100 mm²/s	100 mm²/s	–	50,000 mm²/s
6	Viscosity (-40° C	3000 mm²/s	–	–	–	250 000 mm²/s
7	Flash point	130° C	152° C	140° C	300° C	170° C
8	Interfacial tension	40 mN/m	48 mN/m	>36 mN/m	20–25 mN/m	35 mN/m
9	Dielectric constant ε_r at 20° C	2.25	–	2.66	–	2.23

1. The viscosity of the insulating oil must also be available for the lowest ambient temperature.
2. The pour point must be 10 °C lower than the lowest ambient temperature.

Experience has shown that insulation designs based solely on oil are not a permanent solution, as the particles present in the oil form fibre bridges over time, which leads to breakdowns.

1.2.2　Viscosity and pour point of insulating oils

Figure 1.2 shows the viscosity as a function of temperature for 3 insulating oils. In addition to the insulating oils Shell Diala D hydrofinish and Polyectren D100, the insulating oil Polyectren D12 was also investigated. The measurements were made in the temperature range +20 °C and +80 °C. The values for temperatures <20 °C were not measured. The curves were extrapolated.

The curves show the significantly higher viscosity of Polyectren D100, whose pour point of -50 °C is also at least 10 °C higher than that of Shell Diala D hydrofinish.

The pour point of Shell Diala D hydrofinish was determined at -60 °C and the pour point of Polyectrene D100 at -50 °C.

In practice, the use of Polyectren D100 in Canada has led to several failures of instrument transformers at low temperatures due to the high viscosity and high pour point.

1.2.3 Partial discharge (PD)—inception voltage / field strength as a function of pressure for gas saturated and degassed oils

Figure 1.3 shows the inception voltage and inception field strength of oil insulation for two different sealing systems:

a. Hermetically sealed system with metal bellows (degassed oil)
b. Open system with nitrogen cushion (gas saturated oil)

The curve a. shows an increase in the PD field strength from 20 to 25 kV/mm at a pressure increase from 0.8 to 1.2 bar.

The curve b. shows an increase in the PD field strength from 5 to 20 kV/mm at the same pressure increase from 0.8 to 1.2 bar.

All instrument transformers manufactured in Switzerland between 1965 and 1974 show the behaviour of curve b. because of the adoption of a design with rigid housing and nitrogen cushion. The resulting low partial discharge inception voltage, especially

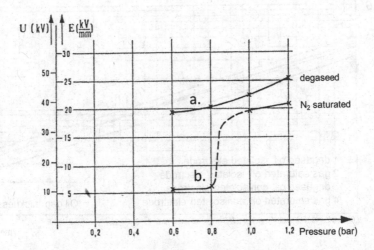

Fig. 1.3 Partial discharge inception as function of pressure for degassed (a.) and gas saturated (b.) insulating oil

in winter, led to numerous instrument transformer failures. About 150 instrument transformers had to be replaced.

1.2.4 Procedure for the assessment and selection of suitable insulating oils

Pure mineral oil, mineral oils with additives, synthetic insulating liquids and natural and synthetic ester liquids are used as liquid insulating agents.

The dielectric strength depends on the examined volume. The thickness of the dimensioned oil gaps plays an essential role. The influence of the length of the oil gap is about 1/10 of the influence of the thickness of the oil gap.

The electrical strength of liquid insulating materials depends on the condition of the insulating liquid. The content of gas, water and solid foreign substances in the insulating liquid is important for this.

For high-quality insulation in instrument transformers, the quality of the processed oil (removal of water, gas, and foreign substances) plays an essential role.

Electrical strength curves have been published by Weidmann, among others [9]. The breakdown voltage of the oil gaps is 22 to 10 kV/mm for an oil section thickness of 1 to 10 mm (see Fig. 1.4).

It is not possible to use large oil gaps for electrical loads for long periods. Experience has shown that fibre bridges can then form which lead to breakdown. From a technical point of view, it is correct to note that free oil gaps are generally a weak point in

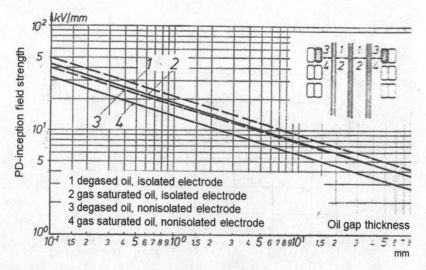

Fig. 1.4 Electrical strength of insulating oils as a function of oil gap thickness in a test arrangement with insulated and non-insulated electrodes

insulation. The authors consider a safety factor of 2 to be necessary. A safety factor of 4 should be observed for oil gaps exposed to longitudinal stress, as can occur during the drying process of the paper. Lower safety factors can be used for insulated electrodes and carefully degassed and particle-free oil.

As an example of a faulty design, free oil paths between the electrode of a bushing and the dome of the transformer should be mentioned. After 3 years of faultless operation, small fibres in the oil built up a bridge between the electrode of the bushing and the dome of the transformer, which led to a breakdown. Barriers had to be installed subsequently to divide the oil path.

1.3   Oil-paper insulation

The outstanding position of the oil-paper insulation system is based on the fact that the paper can be completely impregnated with oil, the oil can dissipate heat from electrical conductors, has a high mechanical stability and a cost-effective insulation can be produced.

1.3.1   The used electrical insulating papers

For the application in the insulation of instrument transformers, the various electrical insulation papers were developed together with the manufacturers. The following applies to all papers used:

- Base material with a sulphate cellulose content of at least 90% in the dry mass
- Ash content $\leq 1\%$.
- Without addition of glues and resins
- Unbleached
- Not calendered

Table 1.2 lists the other parameters of the insulating papers used.

The breakdown voltage and the tanδ are measured with the arrangement shown in Fig. 1.5. Figure 1.6 shows the measured tanδ as a function of temperature. The measurements apply to 4 layers of paper impregnated with insulating oil:

a. Pre-drying under normal pressure at 105 °C for approx. 6 h under atmospheric pressure
b. Evacuation at room temperature to < 13 Pa (0.1 Torr), then 14–16 h at 120 °C
c. Flooding: After cooling to room temperature with Shell Diala D hydrofinish, resting time at a pressure of 13 Pa (0.1 Torr), then aerate with N_2

Table 1.2 Parameters of the insulating papers used.

Thick insulating paper:	80 μm	100 μm	150 μm	150 μm stretchable *	220 μm Crepe paper**
density	0.75 g/cm³	0.60 g/cm³	0.75 g/cm³	0.75 g/cm³	0.55 g/cm³
grammage	62 g/m²	60 g/m²	112 g/m²	112 g/m²	100–150 g/m²
tan δ / 90 °C	3.5×10^{-3}	3.5×10^{-3}	3.5×10^{-3}	3.5×10^{-3}	3.5×10^{-3}
dielectric strength	>19.5 kV	>20 kV	>30 kV	>30 kV	>19 kV
air permeability	40–70 ml/min	>60 ml/min	>42 ml/min	60 ml/min	2–4 ml/min
capillary rise in distilled water:					
lengthwise	10–20 mm	>25 mm	>10 mm	>10 mm	10–40 mm
in transverse direction	10–20 mm	>20 mm	>10 mm	>10 mm	15–40 mm

* Elongation at break in longitudinal direction >6%.
** Elongation at break in longitudinal direction 60%.

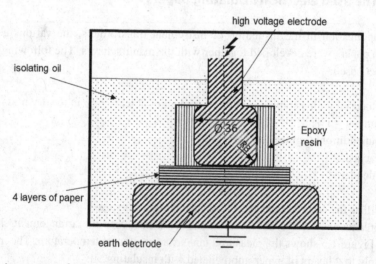

Fig. 1.5 Arrangement for measuring breakdown voltage and tan δ

1.3.2 Application of the insulating papers listed in Table 1.2

The insulating papers listed in Table 1.2 were used in the insulation of the instrument transformers as follows:

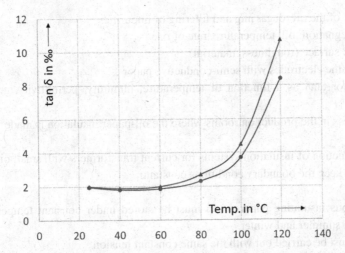

Fig. 1.6 Values measured with the Schering bridge for tanδ as a function of temperature. Measured with 4 layers of impregnated (with Shell Diala D hydrophinish) 100 μm insulating paper with an electrode pressure of 5 kPa. Measuring voltages: ● 190 V (0.5 kV/mm), ▲ 760 V (2.9 kV/mm)

• 80 μm:	Layer winding for voltage transformers, inserts for voltage transformers
• 100 μm:	Electro-insulating paper layers of oil-paper bushing
• 150 μm:	Inserts for connection to core housing for current transformer and the active part for voltage transformer and inserts in the head insulation of current transformers
• 150 μm, stretchable:	Head taping for current transformers
• 220 μm Crepe paper:	Taping of inserts for current and voltage transformers

1.3.3 Influencing parameters of oil-paper technology

Below are some parameters that influence the quality of the oil-paper insulation:

- Structure of the insulation (density in g/m², thickness in μm and choice of insulating paper)
- Distribution of the oil gaps in the insulation
- depth of the oil gap in field direction and volume of the oil gap
- Air permeability of the insulating paper [ml/min], (important for impregnation)
- Relative permittivity ε_r as a function of density and type of cellulose
- Viscosity, temperature, and pressure of the oil

- treatment of the oil (degassing and filtering of impurities)
- For impregnation: oil, temperature, rate of rise
- Electrode surface (roughness, material)
- Covering the electrodes with semi-conductive paper
- Loss factor tanδ as a function of temperature, humidity, permittivity, and type of cellulose
- Cleanliness in the production rooms where the oil-paper insulation is made

For the production of insulation systems for current transformers with small oil gaps it is necessary to keep the boundary conditions constant:

- The various insulating papers used must be stored under constant temperature and humidity; summer and winter.
- Taping must be carried out with the same constant tension.
- The finished active parts must be immediately packed in stable airtight plastic sleeves to avoid contamination.

1.3.4   Measurement of the partial discharge inception and breakdown voltage on insulation models for instrument transformers

The device shown in Fig. 1.7 was used for tests on the electric strength of paper. Since these are comparative measurements on models, no insulated flat electrodes were used but a spherical electrode. All samples were dried in a vacuum and impregnated with Shell Diala D hydrofinish. The glue used is a dextrin glue.

The breakdown voltages on the models, measured with the spherical electrode of 50 mm diameter, resulted:

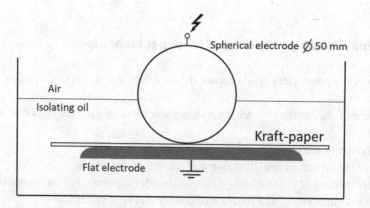

Fig. 1.7 Experimental set-up for measuring breakdown voltage on models

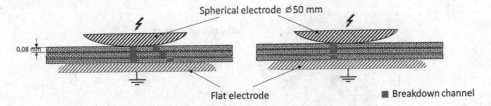

Fig. 1.8 Breakdown channels on two samples with 3 layers of paper

1. For one layer 80 μm Paper: 8.22 kV as average of 4 measurements
2. With 3 layers 80 μm Paper: 16 kV as average of 4 measurements

The samples of 3 layers of paper showed different breakdown channels as shown in Fig. 1.8 The sample in the left-hand figure showed 2 breakdown channels.

1.3.5 Oil-Paper Insulation Systems

1.3.5.1 Types of insulation
The following types of insulation systems are distinguished in instrument transformers:

a. Broadband insulation with capacitive foil inserts for potential grading in the bushing part of current and voltage transformers.
b. Cable insulation overlapped and stepped with very small oil gaps in the head part of the current transformer and in the compact area of the insulation of the high voltage electrode of the voltage transformer.
c. Large and designed oil gaps (2 to 10 mm) and free oil gaps by strip insulation to divide the oil gap between layer insulation and housing in the voltage transformer; as well as for radial stressing of the step of the bushing in the connecting part bushing—torus in current and voltage transformers.
d. Layer insulation of the primary winding in the voltage transformer.

The sectional views of a current transformer (Fig. 1.9) and a voltage transformer (Fig. 1.10) show the location of these types of insulation.

Used materials and their permittivity:

• Oil-paper in broadband bushing:	$\varepsilon_r = 3.5$
• Transformerboard, oil impregnated:	$\varepsilon_r = 3.5$
• Oil-paper hand taping:	$\varepsilon_r = 2.8$
• Mineral oil:	$\varepsilon_r = 2.2$
• Porcelain insulator:	$\varepsilon_r = 6$

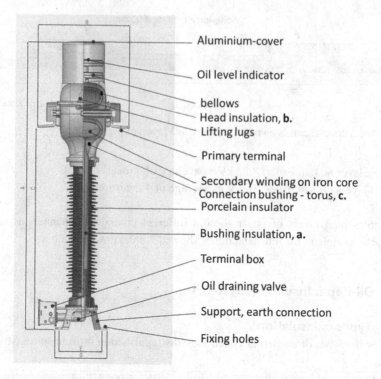

Aluminium-cover

Oil level indicator

bellows
Head insulation, **b.**
Lifting lugs

Primary terminal

Secondary winding on iron core
Connection bushing - torus, **c.**
Porcelain insulator

Bushing insulation, **a.**

Terminal box

Oil draining valve

Support, earth connection

Fixing holes

Fig. 1.9 Sectional view of a current transformer

1.3.5.2 Dimensioning voltage

The insulation in the measuring transformer is designed for a dimensioning voltage U_{DIM} (RMS value), for which the following conditions must be fulfilled:

a. Specified AC test voltage (1 min, RMS value): $U_{\text{DIM}} \geq U_{\text{P}}$
b. Maximum operating voltage $U_m/\sqrt{3}$ (RMS value): $U_{\text{DIM}} \geq 2.6 \cdot U_m/\sqrt{3}$
    to guarantee a service life of 30–40 years.
c. Lightning impulse voltage $\hat{u}$ (peak value): $U_{\text{DIM}} \geq \hat{u}/2.3$
    (The value 2.3 is based on many years of experience)

If the lightning impulse test is chopped with a multiple spark gap, the dimensioning voltage must be checked and possibly increased. Experiences with bushings have shown that 20% is a reasonable increase of the dimensioning voltage.

1.3.5.3 Permissible field strength in the oil gap

The average field strength in oil gaps between solid insulations for U_{DIM} shall not exceed the values shown in Fig. 1.11.

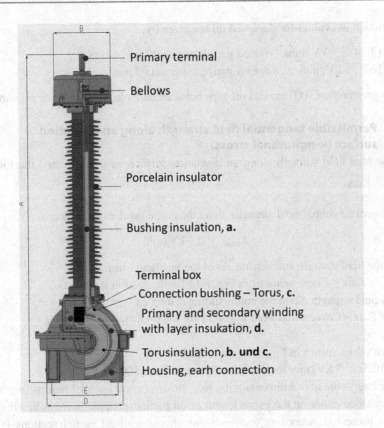

Fig. 1.10 Sectional view of a voltage transformer

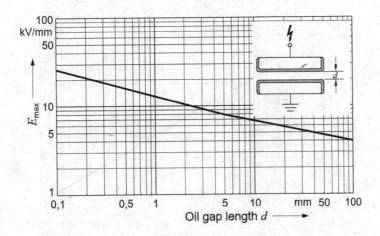

Fig. 1.11 Maximum permissible average field strength in oil gaps for degassed oil with insulated electrodes

The maximum values for degassed oil are given by:

$E_{max} \leq 13.2d^{-0.26}$ kV/mm   for oil gaps between 0.1 and 5 mm.
$E_{max} \leq 10d^{-0,2}$ kV/mm   for oil gaps greater than 5 mm.

Oil gaps greater than 100 mm and oil gaps between bare electrodes are not permitted.

1.3.5.4 Permissible tangential field strength along an insulation surface (longitudinal stress)

The tangential field strength along an insulation surface must not exceed the following values for U_{DIM}:

a. Maximum tangential field strength along the entire insulation surface:

$$E_{tmax} \leq 5.9 \text{ kV/mm}$$

b. Average field strength at a section Δs of the insulation surface
   $1/\Delta s \cdot \int E_t dx \leq$ Curve value from Fig. 1.12 for the section length Δs
c. Mean field strength along the entire sliding distance l
   $1/l \cdot \int E_t dx \leq$ Curve value from Fig. 1.12 for the total sliding distance l

The curve values shown in Fig. 1.12 are given by:
   $E \leq 10.5s^{-0.36}$ kV/mm for distances between 5 and 500 mm.

According to the above dimensioning rule, the average tangential field strengths must not be exceeded either on the entire length or on partial sections. In the following a procedure is presented, according to which this checking of all partial sections is easily possible.

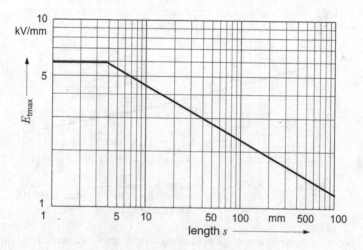

Fig. 1.12 Maximum permissible mean tangential field strength on an insulation surface

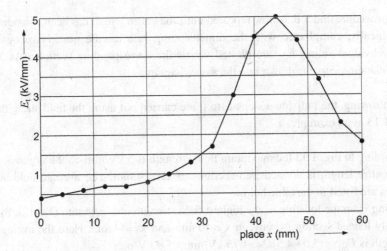

Fig. 1.13 Course of a tangential field strength along a 60 mm long insulation surface

The starting point is the course of the tangential field strength over an insulating material surface as determined from the results of a field calculation. Figure 1.13 shows an example.

Review procedure:

a. First the maximum field strength is checked. According to the dimensioning rule it must not exceed the value of 5.9 kV/mm.

b. The entire length is divided into small sections of equal length Δs starting from the location of maximum field strength:

No	1	2	3	4	5	6	7	8	9	10	11	12	13	14	15	17
Place x mm	0	4	8	12	16	20	24	28	32	36	40	44	48	52	56	60
E_t kV/mm	0.4	0.5	0.6	0.7	0.7	0.8	1.0	1.3	1.7	3.0	4.5	5.0	4.4	3.4	2.3	1.8

c. Starting from the location of highest tangential field strength, the section with the highest average field strength is searched for. This mean field strength E_{m1} must not be greater than the dimensioning specification for the section Δs allows. This means, $E_{m1} \leq 10.5(\Delta s)^{-0.36}$ kV/mm (for $\Delta s \geq 5$ mm), Δs is smaller than 5 mm, E_{m1} must not be greater than 5.9 kV/mm

d. The section selected under c. is now extended by the next section, in the direction that the average field strength over Sect. 2 Δs has the higher value. This average field strength E_{m2} must not be greater than the dimensioning rule for the route $2\Delta s$ allows. This means that $E_{m2} \leq 10.5(2\Delta s)^{-0.36}$ kV/mm

e. The procedure under d. is now repeated until the entire length has been considered.

f. Graphically, compliance with the dimensioning rule can be quickly recognised if the individual values $E_{m\nu}$ are plotted versus the corresponding section $\nu\Delta s$ and the dimensioning curve is drawn into the same diagram.

In the following, the individual steps a. to f. are carried out using the field strength curve of Fig. 1.13 as an example:

a. According to Fig. 1.13 the maximum field strength is 5 kV/mm<5.9 kV/mm

b. The entire length is divided into sections of $\Delta s=4$ mm. The average field strength values are listed in the table below.

c. Starting from the location of the highest field strength at x=44 mm (No.12), the most heavily loaded section is between x=40 mm and x=44 mm. Here the average field strength is $E_{m1}=(5.0+4.5)/2=4.75$ kV/mm<5.9 kV/mm.

d. The distance under c. is extended between x=40 mm and x=48 mm, as the field strength is greater at 48 mm than at 36 mm. The average field strength $E_{m2}=(5.0+4.5+4.4)/3=4.63$ kV/mm. A distance of $2\Delta s=8$ mm would allow an average field strength of 5.0 kV/mm.

e. The section is now increased by 4 mm each time, in the direction of the higher field strength. The table below shows the average field strengths and the permitted values for each section:

ν	Section	$E_{m\nu}$ in kV/mm	$E_{permitted}$ in kV/mm	
0	E_{tmax}	5.0	5.9	☑
1	40 mm–44 mm	4.75	5.9	☑
2	40 mm–48 mm	4.63	4.97	☑
3	40 mm–52 mm	4.33	4.29	!
4	36 mm–52 mm	4.06	3.87	!
5	36 mm–56 mm	3.77	3.57	!
6	36 mm–60 mm	3.49	3.34	!
7	32 mm–60 mm	3.26	3.16	!
8	28 mm–60 mm	3.04	3.02	!
9	24 mm–60 mm	2.84	2.89	☑
10	20 mm–60 mm	2.65	2.78	☑
11	16 mm–60 mm	2.49	2.69	☑
12	12 mm–60 mm	2.35	2.61	☑
13	08 mm–60 mm	2.23	2.53	☑
14	04 mm–60 mm	2.11	2.47	☑
15	Total length	2.01	2.40	☑

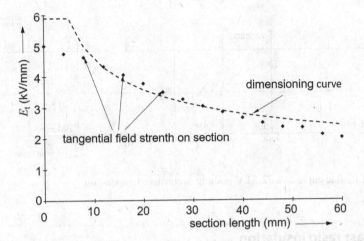

Fig. 1.14 Load on the sections for the example in Fig. 1.13

From this it can be seen that the field strengths E_{m3} to E_{m8} are too high and thus the entire surface is overloaded.

f. Fig. 1.14 shows the load of the individual sections as calculated in point e. together with the dimensioning curve. This figure shows the overloaded sections immediately.

1.3.6 Lifetime of the oil-paper insulation

The oil-paper insulation has a finite life, which depends on the loaded voltage. Under a given continuous stress, the insulation will break down after a certain time. For instrument transformers, the insulation is expected to last at least 40 years. With increasing load voltage, the time until the insulation breaks down decreases. The following applies to the lifetime characteristic curve:

$$U_d/U_0 \sim \left(t_d/t_0\right)^{-1/k}$$

Here, U_0 and t_0 are reference values and k is the lifetime coefficient [10]. In double logarithmic representation, this results in a straight line with a gradient of $-1/k$.

Extensive tests are necessary to determine the lifetime curve of an insulation. Figure 1.15 shows the lifetime curve of a 72.5 kV bushing with oil-paper insulation.

With increasing temperature, the lifetime of the insulation decreases. According to Montsinger, the lifetime is halved if the operating temperature is increased by 8 K [11].

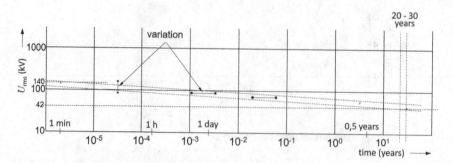

Fig. 1.15 Lifetime curve of a 72.5 kV bushing at constant temperature

1.4   Cast resin insulation

Medium voltage transformers up to 35 kV system voltage are mainly insulated with cast resin.

In principle, two casting resin systems are used, polyurethane and epoxy resin.

1.4.1   Insulation with polyurethane

A frequently used casting resin system is MICARES from ABB Micafil (now part of the special chemicals group Altana). It is a two-component system of resin and hardener. The authors' experience is based on the two resins MICARES 730 and MICARES 721, both mixed with hardener P978. In addition, $MgCaCO_3$ is added as filler to the resin. Table 1.3 shows the properties of these resins. The polyurethane resins can be cured at room temperature but require post-curing at higher temperatures for 7 h to achieve their final hardness.

Casting process: vacuum casting
A vacuum is drawn (< 0.01 MPa) in the resin reservoir to degas the resin. The resin is stirred well to prevent the filler from settling.

Before casting, the mould is heated in the furnace for about 1 h and placed under vacuum of 0.015 MPa for casting.

During casting, resin and hardener are fed to the casting gun under pressure of approx. 0.15 MPa. The stirrer of the resin container must be switched off during this process, otherwise air will be stirred into the resin. In the gun, resin and hardener are mixed in a static mixer and the mixture is filled into the casting mould from below. When the mould is filled with the resin/hardener mixture, an overpressure of about 0.3 MPa is set instead of the vacuum. Then the whole mould is tempered at 40 °C for 4 h and then

Table 1.3 : Polyurethane casting resin systems MICARES.

Resin	MICARES 721	MICARES 730
Hardener	P978	P978
Mixing ratio	2.5:1	5:1
Viscosity at 25 °C (mPa·s)	1200–2200	2500–3000
Setting time (min)	40–80	240–360
Post-curing (h / °C)	7 / 120	7 / 80
Properties after curing:		
Density (g/cm^3)	1.5–1.6	1.7–1.8
Dielectric strength ((kV/mm)	18–24	12–18
tan δ at 23 °C	<0.04	<0.03
tan δ at 70 °C	<0.07	<0.07
Rel. permittivity ε_r at 23 °C	4–5	5.5–6.0
Rel. permittivity ε_r at 70 °C	5–6	8.5–9.2
Tensile strength (N/mm^2, MPa)	50–60	15–25
Flexural strength (N/mm^2, MPa)	85–100	28–30
Glass transition temperature T_g (°C)	95–110	30–40
Thermal expansion coefficient (10^{-6}/K)	35–45	100–120

cured at 60 °C for 10 h. The mould is then cooled down to room temperature. Only then the casting can be removed from the mould.

1.4.2 Insulation with epoxy resin

A second frequently used resin system is epoxy resin. The experience of the authors is based on the epoxy resin Araldite from Huntsman [12]. The system consists of the resin Araldite CW 229–3 and the hardener Aradur HW 229–1. 60% Wollastonite (CaSiO3) is added to the resin and hardener as filler.

The properties of the epoxy resin are listed in Table 1.4.

Casting process: Automatic pressure gelation (ADG)
The resin and hardener are heated to 40 °C in two storage tanks and degassed under vacuum. To prevent the filler from settling, the resin and hardener must be constantly stirred during this process.

Both components are mixed in a third container and degassed at 50–60 °C and a pressure of 3–5 mbar while stirring. Until no more bubbles are present. The gelation time at 50–60 °C is 12 h. Therefore, only as much resin and hardener may be mixed as can be processed during this time.

Table 1.4 : Epoxy resin casting system Araldite 229–3/Araldur 229–1.

Resin	Araldite CW 229–3
Hardener	Aradur HW 229–1
Mixing ratio	1:1
Viscosity at 25 °C (mPa·s)	2000
Setting time (min)	6 at 140 °C
Post-curing (h / °C)	0.5/140 + 10/140
Properties after curing:	
density (g/cm³)	1.8–1.9
Dielectric strength (kV/mm)	18–22
tan δ at 23 °C	0.03
tan δ at 70 °C	0.09
Rel. permittivity ε_r at 23 °C	4.5
Rel. permittivity ε_r at 70 °C	4.7
Tensile strength (N/mm², MPa)	80–90
Flexural strength (N/mm², MPa)	120–130
Glass transition temperature T_g (°C)	110–120
Thermal expansion coefficient (10⁻⁶/K)	27–30

For pouring, the stirrer is switched off and the container is pressurised below 0.3–0.5 MPa. The resin is filled into the mould within 4–6 min via a hose heated to 40 °C.

The casting mould is heated so that curing takes place from top to bottom (see Fig. 1.16). All parts placed in the mould are preheated to 120 °C before assembly.

After a curing time of 30 min the casting can be removed from the mould. The casting mould is then available again for casting the next part. The cast part is post-cured again at 140 °C for 10 h.

Fig. 1.16 Temperature profile of the mould

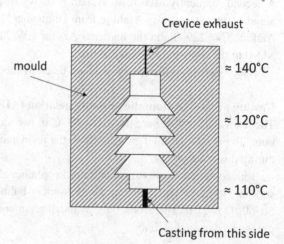

Fig. 1.17 Test pieces for the determination of breakdown voltages

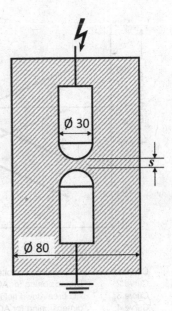

1.4.3 Dimensioning of the cast resin insulation

The insulation of instrument transformers must be able to withstand both alternating voltages and impulse voltages. After tests with cast resin test specimens of 80 mm diameter into which hemispherical electrodes of aluminium with a diameter of 30 mm were cast at distances of s (see Fig. 1.17), the dimensioning curves in Figs. 1.18 and 1.19 were created.

Figure 1.18 shows the breakdown values and the dimensioning curve for the stress with the AC test voltage. The respective breakdown voltages and the breakdown field strengths are shown.

In addition to the AC voltage stress, the lightning impulse voltage load must also be considered. Figure 1.19 shows the breakdown values and dimensioning curves for the lightning impulse voltage 1.2/50 μs.

1.5   SF$_6$ insulation

1.5.1   General information about SF$_6$ as an insulating gas

Sulphur hexafluoride (SF$_6$) has been used as an insulating gas in metal-enclosed switchgear since 1965. Since the 1980s, freestanding instrument transformers with SF$_6$ insulation have also been built. SF$_6$ is a colourless, odourless, non-flammable and non-toxic gas.

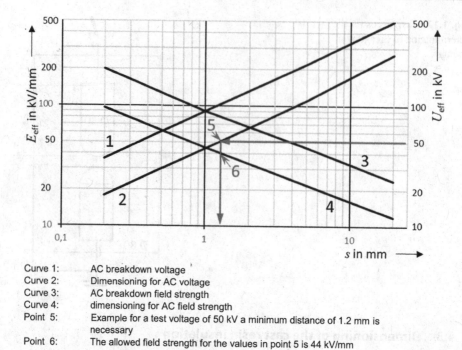

Curve 1:   AC breakdown voltage
Curve 2:   Dimensioning for AC voltage
Curve 3:   AC breakdown field strength
Curve 4:   dimensioning for AC field strength
Point 5:   Example for a test voltage of 50 kV a minimum distance of 1.2 mm is necessary
Point 6:   The allowed field strength for the values in point 5 is 44 kV/mm

Fig. 1.18 Breakdown and dimensioning curves of cast resin insulation for AC voltage. (Curve 1: AC breakdown voltage; Curve 2:Dimensioning for AC voltage; Curve 3:AC breakdown field strength; Curve 4:dimensioning for AC field strength; Point 5:Example for a test voltage of 50 kV a minimum distance of 1.2 mm is necessary; Point 6:The allowed field strength for the values in point 5 is 44 kV/mm)

The quality of new SF_6 gas is defined in IEC 60376 [13], that of used SF_6 gas in IEC 60480 [14] (see Table 1.5). These values must be met when used as an insulating gas in GIS or free-standing instrument transformers.

Table 1.6 shows the values of density and electrical strength of SF_6 in comparison with air.

According to Paschen's law, the breakdown field strength of SF_6 insulation increases with pressure, as shown in Fig. 1.20. In gas-insulated switchgear (GIS) and in SF_6 insulated free-standing instrument transformers, a pressure of 3 to 5 bar is usually used.

SF_6 is gaseous under normal pressure down to a temperature of -63 °C, but at higher pressure this shifts to higher temperatures. From the phase diagram in Fig. 1.21 it can be seen that SF_6 liquefies already at -30 °C under 5 bar pressure. For use at low ambient temperatures, nitrogen is therefore added to SF_6 to reduce the SF_6 partial pressure and thus shift the liquefaction to lower temperatures.

The electrical strength of the SF_6/nitrogen mixture is not greatly reduced even with a larger proportion of nitrogen. An addition of 25% nitrogen only results in a 5% reduction of the electrical strength. A mixture of 70% nitrogen and 30% SF_6 still has a strength of 83% compared to the strength of pure SF_6 (see Fig. 1.22). Thus, the insulating gas can also be used for lower ambient temperatures.

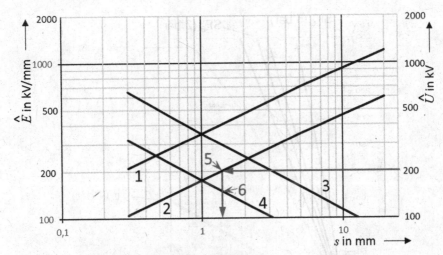

Curve 1:   Breakdown for lightning impulse voltage
Curve 2:   Dimensioning for lightning impulse voltage
Curve 3:   breakdown field strength for lightning impulse voltage
Curve 4:   dimensioning of field strength for lightning impulse voltage
Point 5:   Example for a lightning impulse voltage of 200 kV a minimum distance of
                 1.3 mm is necessary
Point 6:   The allowed field strength for the values in point 5 is 125 kV/mm

Fig. 1.19 Breakdown and dimensioning curves of cast resin insulation for lightning impulse voltage (Curve 1:Breakdown for lightning impulse voltage; Curve 2:Dimensioning for lightning impulse voltage; Curve 3:breakdown field strength for lightning impulse voltage; Curve 4:dimensioning of field strength for lightning impulse voltage; Point 5:Example for a lightning impulse voltage of 200 kV a minimum distance of 1.3 mm is necessary; Point 6:The allowed field strength for the values in point 5 is 125 kV/mm)

Table 1.5 Specifications for SF$_6$ in IEC standards

	IEC 60376 Specification for new SF$_6$	IEC 60480 Specification for used SF$_6$
Content of air or CF$_4$	Max. 1 vol. %	< 3% by volume
Dew point (humidity)	-36 °C at p_a 1 bar < 25 ppm (mass)	-23 °C at < 1 bar p_e (medium voltage) -36 °C at > 1 bar p_e (high voltage)
Percentage oil	< 10 ppm (mass)	< 10 ppm (mass)
HF, SO$_2$ content	< 1 ppm$_v$	< 12 ppm$_v$ SO$_2$ < 50 ppm$_v$ in total

Table 1.6 Physical parameters of SF$_6$ and air

	SF$_6$	Air
Density at 20 °C	6.16 kg/m^3	1.16 kg/m^3
Electrical strength at 1 bar pressure	8.9 kV/mm	2.5 kV/mm

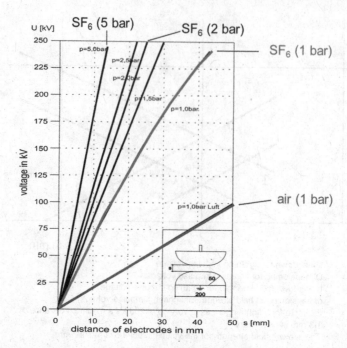

Fig. 1.20 Breakdown voltage of SF$_6$ as a function of pressure

Fig. 1.21 - The SF$_6$ phase diagram

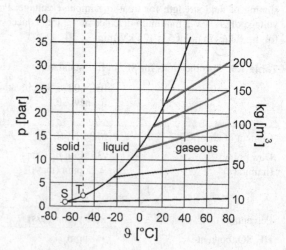

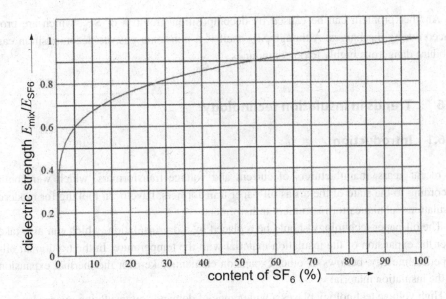

Fig. 1.22 - Breakdown voltage of SF$_6$ / nitrogen mixtures

1.5.2 Influence of electrode roughness on the insulation properties of SF$_6$

When designing an SF$_6$ insulation, special attention must be paid to the surfaces of electrodes. Furthermore, the humidity of the gas and other impurities have a significant influence on the insulation properties.

To trigger a discharge in the electric field, free initial electrons are required, which then trigger further electrons after acceleration in the field, thus leading to the first electron avalanche. The initial electrons are generated by cosmic radiation or are released from the electrode material by high electrical field on the surface. The electrical strength is thus reduced by increased electrical field on rough surfaces. For SF$_6$ insulation, it is therefore important to ensure that all electrodes have smooth surfaces [15].

The emission of initial electrons from the electrode can also be reduced by painting the electrodes with insulating varnish.

1.5.3 Environmental properties of SF$_6$

A disadvantage of SF$_6$ gas is its high global warming potential, which is 22800 times greater than that of carbon dioxide CO_2. For this reason, it is particularly important that SF$_6$ insulated devices are particularly tight. According to current requirements, a maximum leakage rate of 0.5% per year is allowed.

Another problem can be caused by decomposition products of SF_6, which are pro-duced during discharges. Although SF_6 itself is a non-toxic gas, its decomposition can produce dangerous highly toxic compounds.

1.6   Trends in insulation technology

1.6.1   Introduction

In recent years, manufacturers of current and voltage transformers, which were built according to the state of the art as oil-paper transformers, have been looking for modern insulation techniques for instrument transformers.

The oil-paper technology should be replaced by a dry insulation, which can also take over the expansion of the insulation material with the temperature. In the traditional oil-paper technology, bellows or other expansion elements take over the thermal expansion of the insulation material.

High voltage technology places a wide range of demands on insulating materials:

- mechanical properties
- mechanical strength
- thermal stability
- Lifetime
- Electrical properties
  - tan δ
  - Electrical strength
  - Partial discharge free

First investigations with silicone-gel as an insulating material in high-voltage engineer-ing were conducted at the University of Kassel [16]. O. Belz further investigated silicone gel filled with hollow microspheres [17]. They have shown that the silicone gel filled with hollow microspheres is an extremely interesting insulating material and is suitable for use in high voltage instrument transformers.

1.6.2   Insulating material silicone gel with hollow microspheres

O. Belz at the High Voltage Institute of the University of Kassel laid the foundations for the investigation of silicone gel filled with hollow microspheres as an insulating material (Fig. 1.23) [17]. The application of this insulation technique for instrument transformers is described in European Patent EP 2281 294 B1 [18].

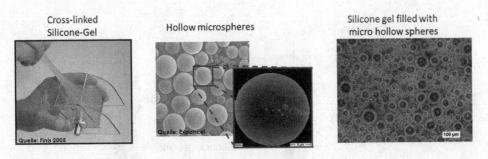

Fig. 1.23 Silicone gel filled with hollow microspheres [17]

There are different types of hollow microspheres on the market, characterised by the size of the spheres, their density, the filling gas, and the material.

Thanks to the admixture of hollow microspheres, which change their volume at different pressures, the silicone gel becomes a compressible material. This allows the use in a wide temperature range.

Unfilled silicone gel is incompressible and has a high coefficient of expansion. This makes it unsuitable as insulation material for instrument transformers.

1.6.3 Dry insulation for instrument transformers

The investigations have shown that oil-paper insulation can be replaced by dry insulation.

The silicone gel filled with hollow microspheres, which is liquid in the non-crosslinked state, can be used as a casting compound. After the material has cross-linked, a solid but compressible mass is formed which sticks to the electrodes and housings.

No additional liquid is required for heat dissipation and no gas can escape in the event of a defect. This means that no environmental measures need to be taken at the installation site.

To use the silicone gel filled with hollow microspheres as an insulating material, however, investigations into electrical and thermal ageing, as well as electrical strength in the geometric dimensions of the insulation are still necessary.

Compared to oil-paper insulated instrument transformers, the silicone-gel insulation and the use of a dry bushing allow the design of an instrument transformer that contains neither oil nor SF_6 as insulating material. When insulating with the filled silicone gel, there is also no need for an expansion bellows (see Fig. 1.24).

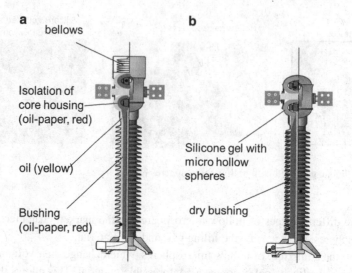

a
bellows

Isolation of
core housing
(oil-paper, red)

oil (yellow)

Bushing
(oil-paper, red)

b

Silicone gel with
micro hollow
spheres

dry bushing

Fig. 1.24 a. Current transformer with oil-paper insulation b. Current transformer with dry insulation

1.7　Porcelain insulators

1.7.1　Material of porcelain insulators

The porcelain insulators are ternary alloys. The three substances are kaolin, quartz, and feldspar.

The composition of the ternary alloy system is described with the help of equilateral triangles. Each point within the triangle corresponds to a composition of the ternary system.

The following illustrations show three equilateral triangles with different inscriptions.

Figure 1.25 shows the composition of a ternary system. Lines are drawn through the point to which an alloy corresponds, parallel to the sides of the triangle. At the points of intersection, the respective concentration of components A, B and C can be read.

Figure 1.26 shows the influence of the mass composition of porcelain on dielectric strength, mechanical strength and temperature resistance. The sides of the triangle are now marked with the basic substances of porcelain, kaolin, quartz, and feldspar [20].

Figure 1.27 shows the concentration range of different refractory ceramics with the ternary concentration diagram [21]. The composition of porcelain for high voltage engineering is limited to a very narrow area on the right side of the triangle.

The composition and properties of ceramic insulating materials can be found in the standard IEC 60672 [22].

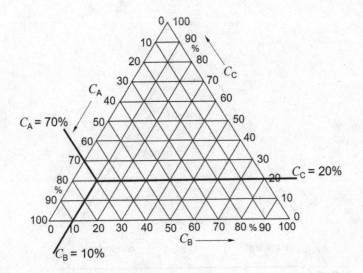

Fig. 1.25 Composition of a ternary system [19]

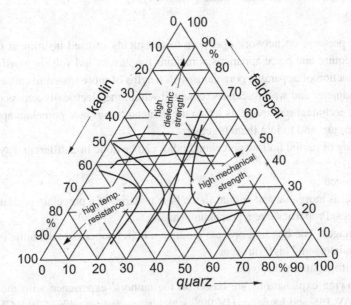

Fig. 1.26 Influence of porcelain composition on dielectric strength, mechanical strength and temperature resistance [20]

1.7.2   Gluing of porcelain insulators

The demand for electrical energy rose steadily between 1970 and 2015. Until 1970 the system voltage of transmission grids was 420 kV / 500 kV, and by 2010 the system voltage reached 800 kV and higher.

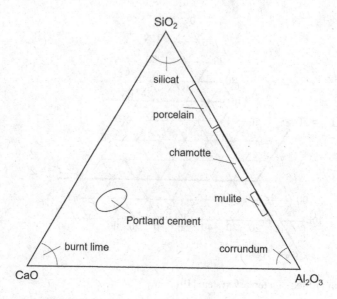

Fig. 1.27 Composition of different refractory ceramics [21]

The cost pressure on network operators forced highly utilised instrument transformers, which require one-piece apparatus porcelains for current and voltage transformers.

The production of apparatus porcelains with lengths of more than 4 m caused difficulties if the diameter and wall thickness remained within the electrically necessary dimensions. These technical requirements led to the demand to produce porcelain apparatus in individual lengths and to join them together.

The joining of partial hollow porcelain bodies can be done in 2 different ways.

a. Glazing together:
   The binder is inorganic and similar to the base material of a porcelain insulator.
   Advantages: Properties like a porcelain insulator
   Disadvantages: The bonding process requires a high temperature as in the production of the insulators and high furnaces.
b. Gluing with organic adhesives:
   The following explanations are based on the authors' experience with the adhesive Araldite AV 138 and hardener HV 998, which were developed by CIBA-GEIGY and are now manufactured and distributed by Huntsman [12].

Araldite AV 138 hardens after addition of hardener HV998 without the application of pressure or heat. It hardens already at temperatures of 5 °C. However, the application of heat accelerates the hardening process considerably and improves the strength of the bond.

Properties of the material:

		Araldite AV138	hardener HV998
		Light beige thixotropic paste	Grey thixotropic paste
Viscosity at 25 °C	Pa s	≈ 960	≈ 385
Viscosity at 50 °C	Pa s	≈ 390	≈ 320
Specific weight at 25 °C	g/cm³	1.6	1.7
Flash point (Pensky-Martens)	°C	110	86

For good adhesion of the adhesive to the glued surface, the porcelain surface must have a minimum roughness depth of $R_a = 10 \pm 4\ \mu m$ ($R_t = 40$ -70 μm). This can be achieved by sandblasting. To remove impurities, the surfaces to be bonded are sanded immediately before bonding with an abrasive cloth of grain size P80 and cleaned with acetone to obtain a grease-free surface. (see Fig. 1.28). Afterwards the glued surface must not be touched with the fingers under any circumstances.

To achieve optimum strength, the mixing ratio of adhesive and hardener is 100 parts by weight Araldite AV 138 and 40 parts by weight hardener HV 998.

The adhesive is applied to both glued surfaces at an ambient temperature of 18 °C to 34 °C with the help of a spatula, pressing it into the pores of the surface. In addition, a bead of adhesive is applied to the lower porcelain part. (see Fig. 1.29).

After the adhesive has been applied, the two parts are joined together (see Fig. 1.30), rotating the upper porcelain section through ± 15° and using a device to align the feet

Fig. 1.28 Preparation of the bonding surfaces by **a.** sanding with abrasive cloth and **b.** cleaning with acetone

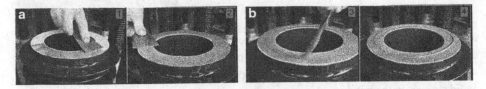

Fig. 1.29 Application of the adhesive on the glued surfaces **a.** with spatula on the upper and lower part and **b.** as an additional adhesive bead on the porcelain lower part

Fig. 1.30 Joining the upper and lower porcelain parts

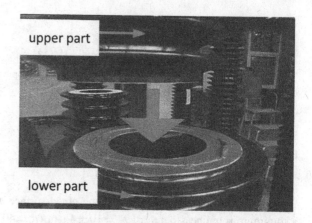

upper part

lower part

Fig. 1.31 Alignment of the feet and head housing of the instrument transformer

and the head housing of the instrument transformer (see Fig. 1.31). A pressure is not necessary. Any adjustment of badly fitting parts by means of pressure is to be avoided. Any necessary adjustment or straightening of the parts must be carried out before bonding to avoid stresses in the bond.

1.7.3 Cementing of porcelain insulators into flanges

The connection between the porcelain insulator and the metallic flange is nowadays usually made by cementing. To achieve a good frictional connection, the ends of the porcelain insulators to be cemented are covered with sand before glazing. The aim is to achieve a mechanical strength that is uniform to that of the porcelain body. The sand should consist of round full grains to avoid overstressing by protruding grains.

Pointed grains lead to overstressing, as shown in Fig. 1.32.

The strength of the cemented porcelain insulator depends on the type of grain used for sanding (Fig. 1.33):

Sanding	classic rounded pointed grain	round grain, uniform size	Round grain, optimum grain size distribution
strength:	100%	110%	140%

The strength of the cemented porcelain insulator is determined by the height of the cementation in relation to the outer diameter of the porcelain. Figure 1.34 shows the breaking strength of the cemented insulator in relation to the ratio of the height of the cementation to the porcelain diameter (h/D) [23]. It is recommended to choose a cementing height of 50% of the outer diameter of the porcelain insulator.

For a good connection between flange and porcelain insulator, the treatment of the cementing surfaces is an essential prerequisite. The area of the insulator to be cemented with the sanding and the flange are cleaned with acetone and then a protective coating

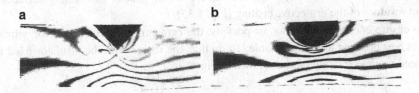

a **b**

Fig. 1.32 Bending stress with sanding of the porcelain with **a**. Pointed grain: The course of the force flow lines shows the overstressing, **b**. Round grain: The overstressing is negligible

Fig. 1.33 Grain size range of a round grain sanding

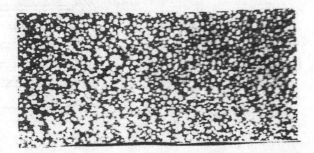

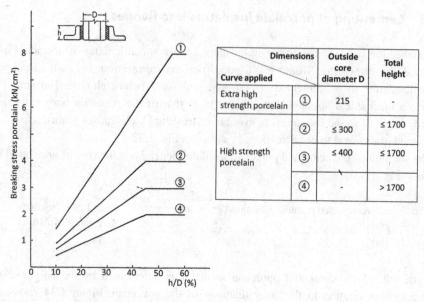

Fig. 1.34 Fracture stress of porcelain as a function of h/D [23]

based on bitumen (Inertol Poxitar) is applied. For this purpose, 100 g Inertol Poxitar is thoroughly mixed with 13 g hardener and 5 g thinner. Then leave the mixture to stand for 30 min, the pot life is 6 h. The protective coating is applied to the porcelain and the flange with a brush.

Special care must be taken to ensure that the O-ring groove and sealing surfaces are free of residues of the protective coating (Fig. 1.35).

To ensure cementing without air pockets, the cementing interface must be vibrated. This is best done on a vibrating table. To do this, the flange with the insulator fixed in it is placed on the vibrating table as shown in Fig. 1.36.

Fig. 1.35 Applying the protection coating to the flange

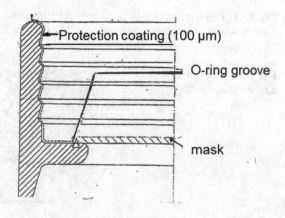

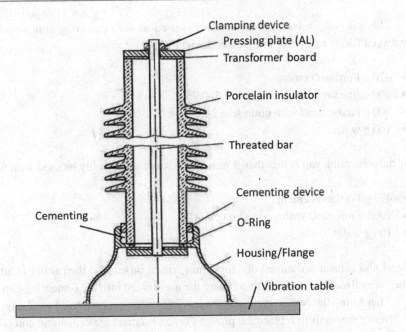

Fig. 1.36 Insulator with instrument transformer housing, with integrated flange, on the vibrating table

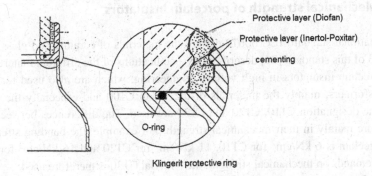

Fig. 1.37 Protecting the porcelain seal with a thin Klingerit ring

The cementing mass is filled into the gap between flange and insulator, and then the material is compacted by vibration. It may be necessary to refill cementing compound afterwards.

The surface is then smoothed, chamfered, and coated with a protective layer of Diofan. This prevents the cementing compound from hardening too quickly, which would lead to cracking. To protect the porcelain sealing surface, a thin Klingerit ring washer is inserted, especially for large insulators (see Fig. 1.37).

The following mixture (coarse sand) is used as the cementing compound for a gap between flange and porcelain greater than 5 mm:

- 470 g Portland cement
- 500 g fine sand with grain size 0.5–0.75
- 500 g coarse sand with grain size 2–3
- 160 g water

If the cementing gap is less than 5 mm, the mixture should only be used with fine sand:

- 470 g Portland cement
- 1000 g fine sand with grain size 0.5–0.75
- 160 g water

Sand and cement are mixed dry for 3 min with a mixer and then water is added. Then they are mixed for another 3 min before the mass is fed into the cementing gap.

After filling the cementing mass, the insulator must not be moved for 7 days.

Newer materials to replace the proven Portland cement are cementing compounds such as Masterflow from BASF. Their advantage is that the insulators can be moved faster.

1.7.4   Mechanical strength of porcelain insulators

The international standard IEC 60672 defines the materials of ceramic and glass insulators. Part 3 of this standard [24] describes the specifications of the different materials.

For porcelain insulators in high voltage engineering, which are also used for instrument transformers, mainly the materials of the group C100 and especially the materials with the designation C110, C120 and C130 are used. The differences between these materials are mainly in their mechanical strength. For example, the bending strength of glazed porcelain is 6 kN/cm^2 for C110, 11 kN/cm^2 for C120 and 16 kN/cm^2 for C130. For high demands on mechanical strength, the material C130 is therefore used.

Table 1.7 shows an extract from the specification of group 100 materials according to the IEC 60672–3 standard.

Table 1.7 : Specification of porcelain insulators of group 100 [24, Table 1].

Properties		Symbol	Subgroup Unit	C110	C120	C130
Flexural Strength	unglazed	σ_{ft}	kN/cm^2	5	9	14
	glazed	σ_{fg}	kN/cm^2	6	11	16
Modulus of Elasticity	Minimum	E	10^3 kN/cm^2	6	-	10

For the mechanical stress of the instrument transformers, as described in chapter 3.2.10, the insulator and in particular the cementing point at the lower end forms the critical area. The material of the insulator must be selected according to the requirements for the load. The weakest point of the insulators is the cementing area, the cementing height should be 50% of the porcelain diameter as described in the previous chapter.

1.8 Composite insulators

1.8.1 Structure of the composite insulators

For free-standing high-voltage apparatus, porcelain insulators are increasingly being replaced by composite insulators. Especially for SF_6 insulated instrument transformers, composite insulators are used to prevent sharp-edged porcelain splinters from endangering devices and persons due to the high pressure in the gas-filled insulator in case of a fault.

The composite insulators consist of a glass fibre reinforced cast resin cylinder with silicone shields (see Fig. 1.38) and metal flanges for fixing to the instrument transformer housings or bases.

The glass fibre tube provides the mechanical strength of the composite insulator. It has a high burst pressure. The insulator does not shatter when bursting. Due to the elasticity of the glass fibre pipe, the composite insulator has a high earthquake resistance.

Fig. 1.38 Design of the composite insulator [25]

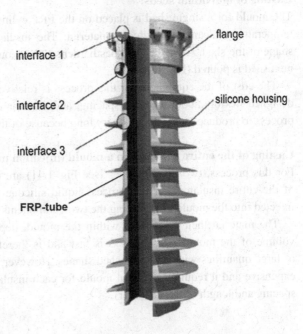

interface 1

interface 2

interface 3

FRP-tube

flange

silicone housing

Fig. 1.39 Hydrophobia of
silicone sheds

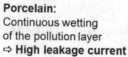

Porcelain: **Silicon elastomer:**
Continuous wetting   Disctrete droplets due to
of the pollution layer   hydrophobic pollution layer
⇨ **High leakage current**   ⇨ **Low leakage current**

The silicone sheds of the insulating tube serve to create the required creepage distance. The hydrophobic properties of the silicone sheds ensure that water rolls off and improve the insulator's pollution performance (see Fig. 1.39). The hydrophobic properties are transferred to any existing layer of dirt. The leakage currents with polluted insulators are significantly lower than with porcelain surfaces. The insulator therefore does not need to be cleaned.

The flange for fastening the composite insulator to the instrument transformer housing is glued directly to the FRP cylinder before the silicone sheds are applied.

The silicone sheds can be applied to the FRP tube using various processes.

Casting of individual sheds:
The mould for a single shed is placed on the FRP cylinder and filled with a RTV (room temperature vulcanizing) silicone material. The insulator rotates to obtain a uniform shape of the shed. After the shed has hardened, the mould is moved on the tube and the next shed is poured (Fig. 1.40).

The cost of the equipment for this process is relatively low, and the insulator can be designed flexibly about the shed spacing, and thus the creepage distance. However, the process to produce an insulator is very long because of the curing of the individual sheds.

Casting of the entire insulator in a mould (injection moulding):
For this process, two mould halves (see Fig. 1.41) are produced with the shed moulds of the entire insulator. For this process liquid silicone rubber (LSR) is used, which is injected into the mould after mixing the two components in a static mixer.

The material then cross-links within the mould, depending on the temperature and volume of the mould. This process is fast and is therefore suitable for the production of large quantities with the same shed shapes. However, the moulds for this process are expensive and it requires a special mould for each insulator (diameter, shed shape, shed spacing and length of the insulator).

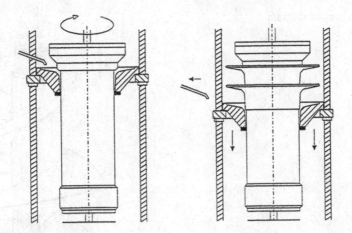

Fig. 1.40 Making the composite insulator by casting individual sheds

Fig. 1.41 Mould for a
composite insulator [26]

Extrusion of the sheds in spiral form:

A special form of composite insulators is created by the spiral extrusion of HTV (high temperature vulcanizing) silicone material. This does not create individual sheds, but a continuous spiral-shaped shed. Any length of insulators with the same shape can be produced (Fig. 1.42).

The flanges of the composite insulator are glued to the FRP tube. This gluing must ensure the mechanical stability of the insulator and guarantee the tightness between the

Fig. 1.42 Schematic diagram
of spiral sheds [27]

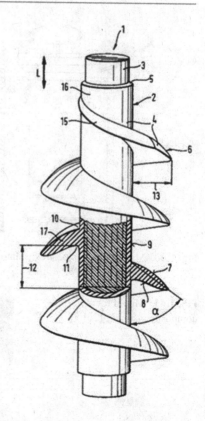

material inside the unit (oil or gas). In many cases this bonding is supported by shrinking
of the flange onto the FRP tube.

1.8.2   Standards and tests

Composite insulators for instrument transformers are described in the standard IEC
61462 [28] "Composite hollow insulators". This standard covers both insulators which
are not under pressure, as with oil-paper insulated instrument transformers, and insulators which are under pressure, as with SF_6 insulated instrument transformers.

The tests are divided into three categories:

Design tests are intended to verify the usability of the design, material, and manufacturing process.

Type tests for a design of composite insulators that have passed the design tests
mainly verify the mechanical characteristics of the insulator.

Sample tests are carried out on a defined number of identical insulators to verify the
characteristics of the insulators, which depend on the manufacturing quality and material. The number of tested insulators depends on the number of insulators produced.

Table 1.8 Loads and pressures relevant to the tests

Bending load on the insulator		
Bending load	Relation	Deformation
Highest load in operation, MML	$= 1.0 \times$ MML	reversible elastic
Elastic limit ("damage limit in IEC 61462)	$> 1.5 \times$ MML	reversible elastic
Load in type test, SML	$= 2.5 \times$ MML	irreversibly plastic
Destructive load	$> 2.5 \times$ MML	irreversibly plastic
Pressure load on the insulator		
Pressure	Relation	Deformation
Highest pressure in operation, MSP	$= 1.0 \times$ MSP	reversible elastic
Pressure for routine tests	$= 2.0 \times$ MSP	reversible elastic
Elastic limit ("damage limit in IEC 61462)	$> 2.0 \times$ MSP	reversible elastic
Pressure during type testing	$= 4.0 \times$ MSP	irreversibly plastic
Specified Internal pressure, SIP	$> 4.0 \times$ MSP	irreversibly plastic

Routine tests are made on each composite insulator to sort out defects in the production process.

Mainly mechanical load tests are described. For this purpose, the necessary mechanical loads and pressures are defined:

SML—"specified mechanical load": bending load specified by the manufacturer for type testing.

MML—"maximum mechanical load": highest mechanical bending load that occurs in operation.

SIP—"specified internal pressure": pressure specified by the manufacturer for type testing.

MSP—"maximum service pressure": maximum pressure during operation.

Table 1.8 summarises the relationship between the different pressures used in the tests.

Further requirements and tests of composite insulators for instrument transformers are contained in the standard IEC 62217 "Polymeric HV insulators for indoor and outdoor use" [29]. The standard IEC 61462 refers in many points to this more general standard.

Bushings

2

2.1   Introduction

The tasks of a bushing are current carrying and connection of high-voltage potentials through a separating earthed surface in different insulating media. (see Fig. 2.1).

The potential difference between the current conductor, the instrument transformer housing, and the environment results in an electric field, which must be controlled by the bushing to ensure that a maximum permissible field strength $\vec{E}_{max}$ on the surface leading to ionisation is not exceeded.

Figure 2.1 shows a schematic bushing principle, which applies to all types of bushings.

The two field strength vectors $\vec{E}_Z$ in longitudinal and $\vec{E}_r$ in radial direction determine the field strength on the surface of the bushing or outside on the porcelain or composite insulator.

If the resulting field strength is too high, the ionisation voltage in air can be exceeded, which is indicated by blue light filaments in the AC test voltage. This process is favoured by humid surfaces in rain.

The two field strengths $\vec{E}_{r\,max}$ and $\vec{E}_{Z\,max}$ in the active part of the bushing along the surface of the insulator, on the high-voltage electrode and near the insulator can be controlled by installing coaxial conductive grading layers or electrodes whose potential distribution is given by the values of the capacitances.

Depending on the number of grading layers or electrodes, smaller values for $\vec{E}_r$ and $\vec{E}_Z$ can be achieved. Figure 2.2 shows the distribution of the potential lines at a capacitor bushing (number of grading layers >20 depending on the system voltage).

© Springer Fachmedien Wiesbaden GmbH, part of Springer Nature 2022
R. Minkner and J. Schmid, *The Technology of Instrument Transformers*,
https://doi.org/10.1007/978-3-658-34863-2_2

Fig. 2.1 Bushing principle

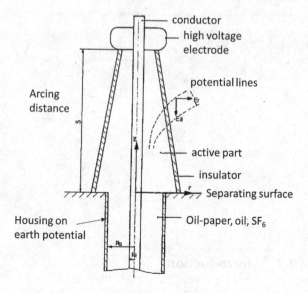

conductor
high voltage
electrode

potential lines

active part

insulator

Separating surface

Oil-paper, oil, SF₆

Arcing
distance

Housing on
earth potential

Fig. 2.2 Potential lines
of a fine graded condenser
bushing using the example of a
transformer bushing

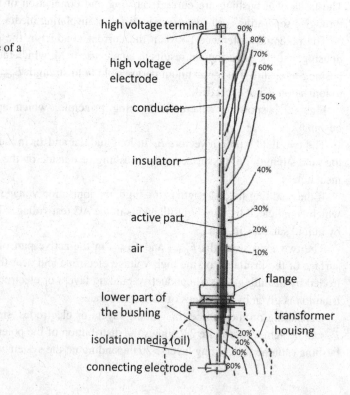

high voltage terminal

high voltage
electrode

conductor

insulatorr

active part

air

lower part of
the bushing

isolation media (oil)

connecting electrode

flange

transformer
houisng

In this design principle, the grading layers in the active part control the electrical field of the bushing on both sides of the separation plane. On the air side, there is a uniform potential reduction from the high-voltage electrode to the flange along the porcelain insulator. The field strength at the high voltage electrode, which is essential for the guaranteed RIV values, can also be influenced.

The electrical strength of the insulating oil, oil-paper insulation and SF_6 gas allows higher surface gradients along the bottom of the bushing and the grading layers force a constant longitudinal gradient. The field then changes into the given electric field of the active parts: voltage transformer layer winding, or metallic shell at ground potential with measuring cores in current transformers.

2.2 Oil-Paper (OIP) Bushings

2.2.1 Fine and Coarse Graded Bushings in Instrument Transformers

Each high voltage instrument transformer has an integrated bushing, which is designed according to the principle of the transformer bushing. The standard IEC 60137 [30] applies to bushings in general and to transformer bushings in particular.

For the design of bushings for instrument transformers, basic characteristics are given for different types of bushings, such as partial discharge values, which are prescribed in the instrument transformer standards.

In principle, there are two different designs which are used for the instrument transformers:

a) Separate machine production of bushings with a step-shaped connecting part to the active part: the voltage transformer winding or the metallic shell with the current transformer cores specified by the customer (see Fig. 2.3)

Requirements for the separately manufactured instrument transformer bushing:

- Partial discharge values <1 pC up to the test voltage with a probability of <0.1%.
- Easy connection of the bushing to the active part of the instrument transformer
- Optimal utilisation of the bushing insulation
- Easy adaptability to the requirements of the customers

b) Joint production of insulation of bushing and active part: winding in the case of voltage transformers, or metallic shell with toroidal cores in the case of current transformers. (see Fig. 2.4)

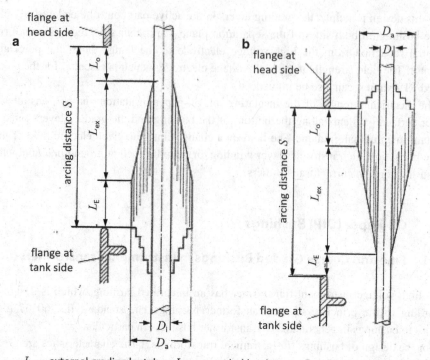

L_{ex} – external grading lentgh, L_o – ungraded lenth top, L_E – length of earth layer

Fig. 2.3 Instrument transformer bushings **a** Voltage transformer **b** Current Transformer

Typical for the design principle according to b) is a coarse grading, with ring-shaped electrodes on the bushing adapted to the field. The metallic grading layers connected to the electrodes are placed around the entire active part in current transformers. The increased capacitance between the control layers increases the capacitive current and thus promotes the triggering of partial discharges.

The basis for the field strength $\vec{E}_r$ between the layers $\vec{E}_r = f(d)$ is based on tests described below.

2.2.2 The Determination of Field Strength Between the Grading Layers of Fine Graded Condenser Bushings

2.2.2.1 Development Tests

It was known from Paschen's law for gases and oil-paper technology that the breakdown voltage or the breakdown strength for AC and impulse voltages is a function of the insulation thickness.

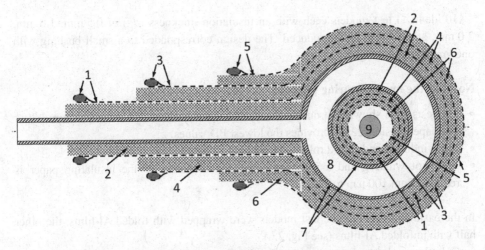

1. Potential electrode (ring) with metal layer (1/3 high voltage potential)
2. Insulation
3. Potential electrode (ring) with metal layer (2/3 high voltage potential)
4. Insulation
5. Potential electrode (ring) with metal layer (high voltage potential)
6. Insulation
7. Core shell
8. Measuring cores
9. Primary current

Fig. 2.4 Coarse grading of a bushing in a live tank current transformer

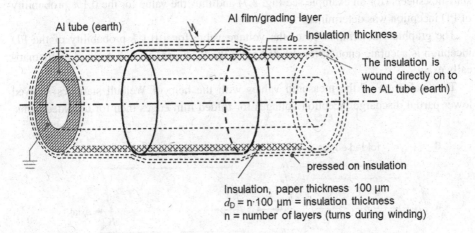

Al tube (earth)

Al film/grading layer

d_D Insulation thickness

The insulation is
wound directly on to
the AL tube (earth)

pressed on insulation

Insulation, paper thickness 100 µm
$d_D = n \cdot 100$ µm = insulation thickness
n = number of layers (turns during winding)

Fig. 2.5 Design of the test models

10 identical test models each with an insulation thickness (d_D) of 0.5 mm, 1.0 mm, 2.0 mm, and 5.0 mm were produced. The design corresponded to a small bushing with one grading layer (Fig. 2.5).

Notes on the manufacturing process:

- The tests have shown that during the winding process a certain roller pressure $\pm 30\%$ and paper tension $\pm 30\%$ provides the lowest PD values.
- The winding mandrel and rollers are heated to 120° C.
- With hot winding and a roller pressure, the thickness of the insulating paper is reduced from 100 µm to 76 µm.

In the basic tests, half of the test models were wrapped with folded Al-films, the other half with unfolded Al-films (see Fig. 2.6).

Only the axial cut edges are critical for the films, as the radial cut edges overlap.

The measurement of the PD values of the test samples showed that the PD inception values of the test samples with folded grading films (solution a) were lower than with unfolded films (solution b). Solution b) was chosen as standard. The production technique with unfolded control foils is also simpler. The lower PD inception stresses with folded films are due to the slightly larger oil gap at the edge of the film. The assessment of both versions a) and b) can change due to different manufacturing methods.

2.2.2.2 The Evaluation of PD Measurements

The five voltage values for the partial discharge inception were entered into the Weilbull statistics sheets (for an example see Fig. 2.7) and thus the value for the 0.1% probability of PD inception was determined.

The graphical determination of the voltage value for a 0.1% probability of the PD inception is accurate enough for the design of the bushings, even if it is not mathematically exact.

The evaluation of the measured values with the help of Weibull statistics showed lower partial discharge inception voltages for folded film edges than for unfolded films.

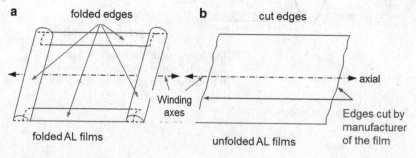

Fig. 2.6 Style of inserted films

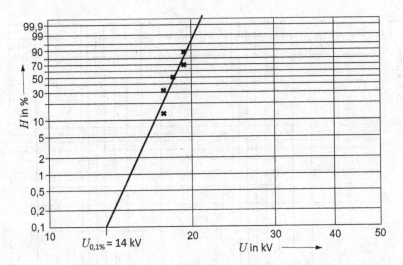

Fig. 2.7 The Weilbull diagram for the example 0.5 mm dielectric thickness with film edges not folded

The reason for this is the larger oil gap at the edge of the film when the film is folded over.

The electrical strength for the design at test voltage $\vec{E}_{rP}$ for a PD inception voltage with PD < 1 pC is given by $|\vec{E}_{rP}| = \frac{U_{PD}(<1pC)kV_{rms}}{0.5mm}$.

In general, the radial field strength is given by the expression $\vec{E}_r = \frac{U}{r \cdot \ln {^{r_a}/_{r_i}}}\vec{e}_r$.

The inner radius r_i and the outer radius r_a are fixed for a bushing, and the field strength $\vec{E}_r \sim {^1/_r}$ decreases from the inside to the outside.

With the experimentally determined relationship ($\vec{E}_r = f(d)$, see Fig. 2.8) it is possible to vary the distance of the grading length and to dimension the longitudinal grading of the upper air part and the lower oil part of the bushing optimally.

The axial stress distribution is determined by the radial voltage of the grading layers. The radial voltage is given by the capacitive distribution by means of the grading layers, conductor tube and earth layer/flange.

Figure 2.8 shows the values of the measurements determined by Weibull statistics (according to Table 2.1 and Fig. 2.7) together with a design curve for bushings. The low values of the design curve consider the reduction of the lifetime at increased operating temperatures according to the Montsinger rule [11].

The maximum permissible field strength $\vec{E}_r$ in the radial direction between two grading layers as a function of the oil-paper dielectric thickness between the layers (Fig. 2.9) is given by the equation $\vec{E}_r = 9.5 \cdot d^{-0.63} \cdot \vec{e}_r \frac{kV_{rms}}{mm}$. This design curve is shown in Fig. 2.8.

With the help of a computer programme the positions of the grading films are calculated, considering the field strengths according to the design curve (Fig. 2.8, curve c).

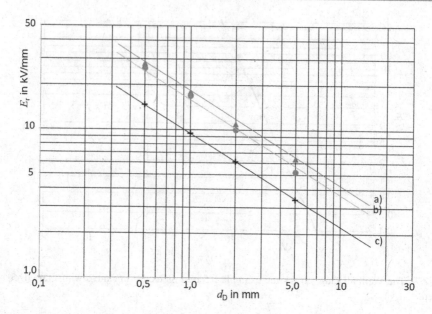

Fig. 2.8 Radial field strength $\vec{E}_r$ as a function of dielectric thickness d_D for 0.1% probability of PD inception a) ▲ measured values for unfolded film edges b) ● measured values for folded film edges c)+design curve

In addition, the uniform distribution of the potential in axial direction is considered as a boundary condition. The distances of the films are determined in such a way that the radial field strength between two films reaches as high a value as possible but does not exceed the design curve (Fig. 2.8, curve c). However, a minimum spacing of 6 paper thicknesses is maintained to take account of any imperfections in the paper. The grading films should have a constant length in the axial direction.

The method of maximum and minimum values of several variables and the Monte Carlo method have proven to be an excellent tool for the simultaneous optimisation of different values [31].

In addition to the electrical parameters, the mechanical values must also be considered when designing the bushing.

2.2.3 The Connection Area

The machine-made oil-paper bushing for a current or voltage transformer must be connected to the active parts; current transformer core housing for the measuring and protective ring cores or to the layer winding of the voltage transformer (see also Sect. 3.2.8.2).

The design of the bushing connection area is based on the technology of the bushing lower part (steps see Fig. 2.10).

Table 2.1 Partial discharge inception voltage measured on 5 samples each with folded and unfolded films, and voltage determined according to Weibull statistics for 0.1% probability of partial discharge

a) folded foil edges

d_D	1	2	3	4	5	0.1% Value
0.5 mm	18.4 kV	17.3 kV	16.5 kV	18.3 kV	17.4 kV	13 kV
1 mm	24.0 kV	24.9 kV	22.8 kV	26.0 kV	23.7 kV	17 kV
2 mm	32.6 kV	33.0 kV	33.1 kV	34.7 kV	–	20 kV
5 mm	49.0 kV	27.4 kV	30.0 kV	45.7 kV	–	24 kV

b) Non-folded film edges

d_D	1	2	3	4	5	0.1% Value
0.5 mm	19.2 kV	19.4 kV	18.4 kV	17.5 kV	17.5 kV	14 kV
1 mm	24.7 kV	24.0 kV	25.3 kV	26.8 kV	24.8 kV	18 kV
2 mm	40.0 kV	41.4 kV	31.9 kV	35.0 kV	38.0 kV	22 kV
5 mm	44.9 kV	45.1 kV	44.0 kV	39.0 kV	–	32 kV

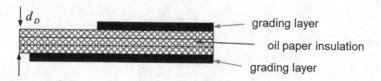

Fig. 2.9 Arrangement of grading layers in the bushing

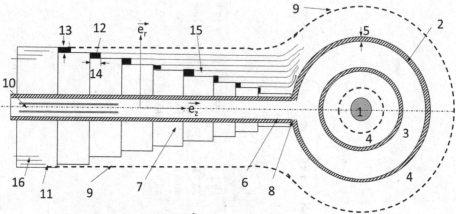

1. Conductor on high voltage
2. Core shell at earth potential
3. Measuring and protective cores at earth potential
4. Isolation
5. Wall thickness of the core shell
6. Support tube of the bushing at earth potential
7. High voltage bushing manufactured separately
8. Welded support tube of the bushing
9. Metallic high voltage coating, connected to the head housing
10. Test leads to the cores
11. Connection of high-voltage coating with the last foil of the fine graded capacitor bushing
12. Oil gap between the bushing and taped insulation
13. Oil gap depth d_T
14. Oil gap width d_B
15. Strip insulation attached to the steps
16. grading layers of the bushing

Fig. 2.10 Connection area bushing to active part

The connection is made with special paper inserts, which are glued to the steps of the bushing. The insulation of this transition is made by hand.

Figure 2.10 shows the connecting part bushing and active part of the current transformer.

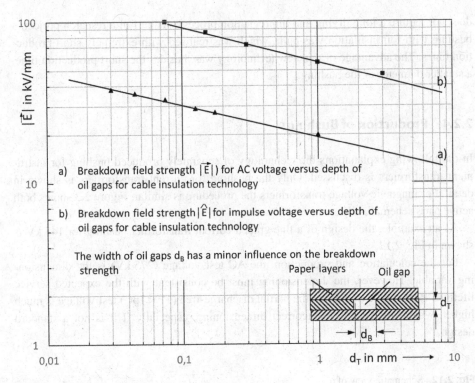

Fig. 2.11 Breakdown field strength in oil gaps

The resulting oil gaps (position 12, 13 and 14 in Fig. 2.10) are dimensioned with the help of the curves in Fig. 2.11.

The field strength vectors for the AC test voltage $\vec{E}_{rms}(r.z)$ and the specified impulse voltage level (BIL) $\hat{E}(r.z)$ between the high voltage layer 9 and the support tube 6 must be calculated for the oil gap depth d_T and oil gap width d_B.

The gaps are created when the strip insulation is taped to the steps and by shrinking of the paper of the bushing and the strip insulation during the drying process of the insulation before impregnation with oil.

The calculated field strength $\vec{E}(r,z)$ must be broken down into the two components $\left|\vec{E}_r\right| \cdot \vec{e}_r$ and $\left|\vec{E}_z\right| \cdot \vec{e}_z$ Since the electric field for the individual stages is approximately cylindrical, the stress in the z-direction can be neglected.

The AC field strengths $\left|\vec{E}_{rms}\right| \cdot \vec{e}_r$ and the impulse voltage field strengths $\left|\hat{\vec{E}}_r\right| \cdot \vec{e}_r$ must be calculated. The values shall be smaller than the values in Fig. 2.11 with a safety factor of 1.4 to 2. The field strengths shall be calculated with $\varepsilon_r(\text{oil}) = 2.2$ and $\varepsilon_r(\text{paper}) = 3.5$, or according to other associated material values.

The oil paper bushing has proven to be an excellent and reliable element for instrument transformers. At the same time, tests were made with plastic film insulation, which does not require a long drying and impregnation process. As part of a research project, a bushing with foil insulation was built, which was realised with extremely short production times. The advantage of this new technology was, besides the short production time, a smaller diameter of the bushing.

2.2.4 Production of Bushings

In the following explanations the technology of separately produced bushing for instrument transformers is discussed. Only the bushing for current transformers is shown in detail. For magnetic voltage transformers the procedure is similar. Figure 2.3 shows both active parts schematically.

As an example, the design of a fine graded current transformer bushing for 145 kV is shown in Fig. 2.12.

For the calculation and production, the AC test voltage (275 kV) is the dimensioning variable. However, the dimensioning must be consistent with the expected service lifetime. It should be noted that the partial discharge-free (TE < 1pC) test voltage is much higher, which often leads to incorrect dimensioning, especially if it is not a standard design.

Fig. 2.12 Schematic view of a current transformer bushing

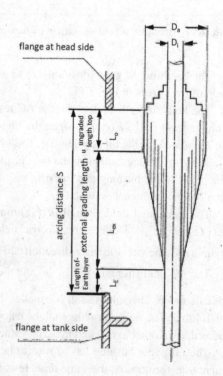

Fig. 2.13 Finished bushing (for current transformer) for 275 kV AC test voltage

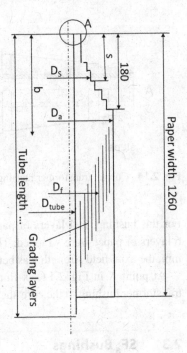

Dimensions of the bushing in Fig. 2.13:

- D_{tube}: outer diameter of the conductor tube 40 mm
- D_a: outer diameter of the bushing 85 mm ± 1 mm
- film width: 350 mm
- Film thickness: 6 μm
- Number of aluminium films: 37
- Grading length 640 mm
- striking distance 1200 mm
- Dielectric: 100 μm Insulating paper pressed and hot wound, gives a thickness of the paper of 76 μm

The production of the instrument transformer bushing is based on 2 calculation programs:

a) Placing of the grading films
b) Calculation of the radial and axial field strengths at each inserted film for the test voltage of 275 kV

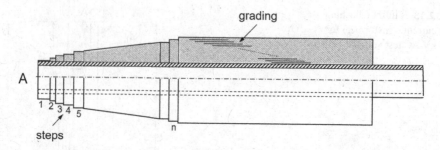

Fig. 2.14   Current transformer bushing 145 kV/275 kV

For this bushing, 294 layers of paper were needed. Between the films, between 14 and 6 layers of paper were wrapped. The radial field strength was between 8.8 and 15.4 kV/mm, the axial field strength was between 0.24 and 0.64 kV/mm.

At point A in Fig. 2.14 six tinned copper strips are inserted to connect the current transformer bushing to the core shell for fast transient voltages.

2.3   SF$_6$ Bushings

For instrument transformers with SF$_6$ insulation, bushings with SF$_6$ insulation are also used. As gas-insulated instrument transformers and bushings are operated at a pressure of several atmospheres, composite insulators are used in most cases instead of porcelain insulators for safety reasons.

To avoid high field strengths at the flanges of the bushings, the potential is controlled along the bushing. For SF$_6$ insulated bushings, simple electrodes are used for lower voltages up to about 170 kV. For voltages above 245 kV capacitive grading is mostly used.

2.3.1   Field Grading with Electrodes

To limit the field strengths at the flanges of the bushing, electrodes are installed in the bushing as shown in Fig. 2.15. The electrodes must be dimensioned so that:

- The field strengths at the electrode surface within the bushing do not lead to discharges,
- The field strengths at the flanges outside the bushing are reduced so that external discharges do not occur.
- The field strength at the surface of the insulating material is limited so that erosion of the silicone sheds does not occur.

Fig. 2.15 Electrode control in SF$_6$ insulated bushing [32]

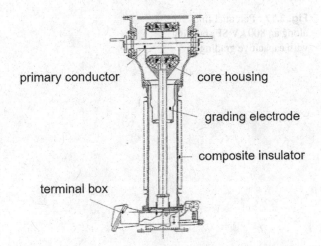

primary conductor

core housing

grading electrode

composite insulator

terminal box

Fig. 2.16 SF$_6$ insulated current transformer with capacitive potential grading

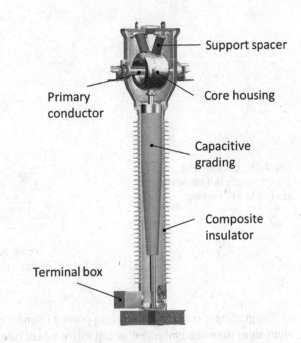

Support spacer

Core housing

Primary conductor

Capacitive grading

Composite insulator

Terminal box

2.3.2 Capacitive Fine Grading for SF$_6$ Bushings

Bushings for voltages of 245 kV and higher are usually not realised with an electrode for field grading but are equipped with a capacitive fine grading which controls the potential along the bushing length.

Figure 2.16 shows an SF$_6$ insulated current transformer with capacitive potential grading in the bushing [25].

Fig. 2.17 Potential lines
along an 800 kV SF$_6$ bushing
with capacitive grading

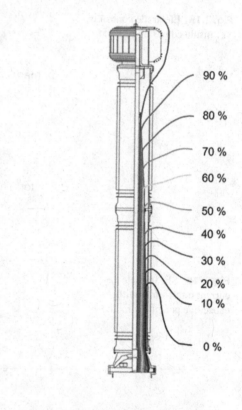

90 %

80 %

70 %

60 %

50 %

40 %

30 %

20 %

10 %

0 %

Fig. 2.18 Detail of the
grading layers in a capacitive
grading in SF$_6$ bushings

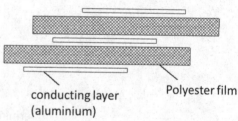

conducting layer
(aluminium)

Polyester film

The grading is realised by a self-supporting cylinder made of polyester foils, in which
aluminium films are embedded as capacitive layers (see Fig. 2.18). These coatings form
a capacitive distribution of the voltage along the bushing and thus control the potential
distribution. Figure 2.17 shows the potential lines on an 800 kV bushing using capacitive
grading.

Current Measurement

3

3.1 Introduction and Standardisation

Conventional current transformers based on the inductive transformer principle are still used in most cases as instrument transformers for current measurement. In addition to the precise transmission of the primary current to the secondary side, they also transmit power of up to 100 VA, depending on the type of protection and measuring devices connected. The characteristics, requirements and different designs of conventional current transformers are discussed in more detail in Sect. 3.2. Many theoretical backgrounds and technical aspects about conventional current transformers are given in the book "Die Messwandler" by Rudolf Bauer [1].

With the development of protection and measuring devices towards purely electronic and digital devices, the demand on the power to be transmitted by the current transformer became smaller and smaller. This made it possible to use current transformers of other principles, such as low-power current transformers, air-core coils, or optical current transformers. These new, non-conventional current transformers are discussed in Sects. 3.3 to 3.5.

The first IEC standard for instrument transformers appeared as early as 1931 as IEC 44 "Recommendations for Instrument Transformers". Today, the IEC 61869 "Instrument Transformers" series of standards applies to all instrument transformers. The requirements for conventional current transformers are given in Parts 1 and 2 of this series of standards. Part 1 [33] contains general requirements for all instrument transformers and Part 2 [34] additional requirements for conventional inductive current transformers. A newly published Part 6 of the series of standards [35] describes additional requirements for transformers with low output power, including electronic instrument transformers containing active electronics, and passive instrument transformers without active electronics. There is a special part of the standard series with additional requirements for

© Springer Fachmedien Wiesbaden GmbH, part of Springer Nature 2022
R. Minkner and J. Schmid, *The Technology of Instrument Transformers*,
https://doi.org/10.1007/978-3-658-34863-2_3

each different measuring principle. Part 8 [36] deals with electronic current transformers containing active elements, such as optical transformers (see Sect. 3.5), or current transformers with amplifiers. Part 10 [37] deals with passive low-power current transformers without active elements, such as iron-core current transformers with shunt resistor (see Sect. 3.3), or air-core current transformers (see Sect. 3.4). A possible digital output of the current transformer is described in Part 9 of the series of standards [38] and is based on the transmission of sampled measured values of the communication standard IEC 61850 [39]. Figure 3.1 shows an overview of the parts of the IEC 61869 series of standards

PRODUCT FAMILY STANDARDS		PRODUCT STANDARD	Title	German standard
IEC 61869-1 GENERAL REQUIREMENTS FOR INSTRUMENT TRANSFORMERS German Standard: DIN EN 61869-1 VDE 0414-9-1		IEC 61869-2	ADDITIONAL REQUIREMENTS FOR CURRENT TRANSFORMERS	DIN EN 61869-2 VDE 0414-9-2
		IEC 61869-3	ADDITIONAL REQUIREMENTS FOR INDUCTIVE VOLTAGE TRANSFORMERS	DIN EN 61869-3 VDE 0414-9-3
		IEC 61869-4	ADDITIONAL REQUIREMENTS FOR COMBINED TRANSFORMERS	DIN EN 61869-4 VDE 0414-9-4
		IEC 61869-5	ADDITIONAL REQUIREMENTS FOR CAPACITIVE VOLTAGE TRANSFORMERS	DIN EN 61869-5 VDE 0414-9-5
	IEC 61869-6 ADDITIONAL GENERAL REQUIREMENTS FOR LOW-POWER INSTRUMENT TRANSFORMERS German Standard: DIN EN 61869-6 VDE 0414-9-6	IEC 61869-7	ADDITIONAL REQUIREMENTS FOR ELECTRONIC VOLTAGE TRANSFORMERS	DIN EN 61869-7 VDE 0414-9-7
		IEC 61869-8	ADDITIONAL REQUIREMENTS FOR ELECTRONIC CURRENT TRANSFORMERS	DIN EN 61869-8 VDE 0414-9-8
		IEC 61869-9	DIGITAL INTERFACE FOR INSTRUMENT TRANSFORMERS	DIN EN 61869-9 VDE 0414-9-9
		IEC 61869-10	ADDITIONAL REQUIREMENTS FOR LOW-POWER PASSIVE CURRENT TRANSFORMERS	DIN EN 61869-10 VDE 0414-9-10
		IEC 61869-11	ADDITIONAL REQUIREMENTS FOR LOW-POWER PASSIVE VOLTAGE TRANSFORMERS	DIN EN 61869-11 VDE 0414-9-11
		IEC 61869-12	ADDITIONAL REQUIREMENTS FOR COMBINED ELECTRONIC INSTRUMENT TRANSFORMER OR COMBINED LOW-POWER PASSIVE INSTRUMENT TRANSFORMERS	DIN EN 61869-12 VDE 0414-9-12
		IEC 61869-13	STAND ALONE MERGING UNIT	DIN EN 61869-13 VDE 0414-9-13
		IEC 61869-14	ADDITIONAL REQUIREMENTS FOR CURRENT TRANSFORMERS FOR DC APPLICATIONS	DIN EN 61869-14 VDE 0414-9-14
		IEC 61869-15	ADDITIONAL REQUIREMENTS FOR VOLTAGE TRANSFORMERS FOR DC APPLICATIONS	DIN EN 61869-15 VDE 0414-9-15

Fig. 3.1 IEC 61869 series of standards with parts 1 to 15

together with the corresponding German standards DIN EN 61869 with parts 1 to 15, which were also included in the VDE regulations as VDE regulation VDE 0414-9 parts 1 to 15.

3.2 Conventional Current Transformers with Iron Core

3.2.1 The Inductive Principle of the Conventional Current Transformer

3.2.1.1 The Basic Principle

A conventional current transformer is an inductive transformer in which the secondary winding is almost short-circuited. The secondary winding is wound around the iron core. In most cases, the primary winding is only formed by one conductor carrying the primary current through the opening in the iron core. For use with low primary currents, however, the primary winding can also contain several turns to improve the properties of the current transformer.

Ideally the secondary current i_S is proportional to the primary current i_P and the ratio of the number of turns of the primary winding N_P and the secondary winding N_S (see Fig. 3.2). The secondary current can be calculated as $i_S = i_P\, N_P/N_S$. The accuracy of the current transformer depends on the size and material of the iron core and the burden connected to the secondary terminals.

3.2.1.2 Core Materials and Their Magnetisation Curves

The magnetisation curve in Fig. 3.3 shows the relationship between the magnetising current and the magnetic flux density in the iron core. As long as the core is not saturated, the magnetising current remains very small and therefore the error of the current transformer remains small. With the saturation of the current transformer core, the magnetising current and therefore the error of the current transformer increases strongly.

Depending on the application and the requirements on the accuracy of the current transformer, silicon iron cores with a saturation induction of 1.8 Vs/m^2 or nickel iron cores with a saturation induction of 0.8 Vs/m^2 are used.

Fig. 3.2 Basic principle of the inductive current transformer

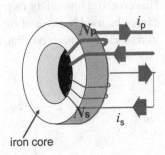

iron core

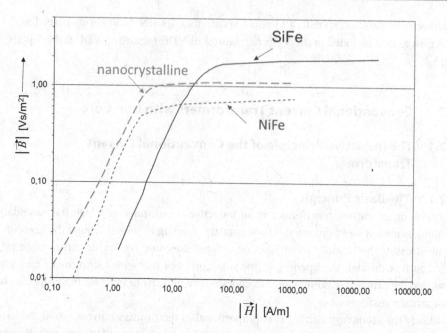

Fig. 3.3 Typical magnetisation curve of a current transformer core of silicon iron (SiFe), nickel iron (NiFe) and nanocrystalline material

Silicon iron cores are used for protective applications and simple measurement applications. They are suitable for this because of their later saturation and higher saturation flux. For some measurement applications where, high accuracy is required (0.2 or 0.1%), nickel–iron cores are used because of the lower magnetising currents. However, these are about 10 times more expensive than silicon iron cores and are therefore only used if the accuracy requirements make it necessary. Since the nickel–iron core goes into saturation at lower induction, a small safety factor FS (see Sects. 3.2.3) can only be achieved with these cores.

More recently, nanocrystalline cores have also been used. Figure 3.3 shows that these cores require an even smaller magnetising current than nickel–iron cores. In many cases they can be used to replace the expensive nickel–iron cores for measurement purposes. However, if a low safety factor FS is required (see Sect. 3.2.3), the nanocrystalline core cannot be used because of the higher saturation flux density of 1.2 Vs/m². Both nickel–iron cores and nanocrystalline cores require a protection for mechanical stability.

3.2.2   Errors of the Inductive Current Transformer and Their Influencing Factors

3.2.2.1 The Equivalent Circuit Diagram of the Current Transformer

Figure 3.4 shows the equivalent circuit diagram of an inductive current transformer reduced to the secondary side. The primary winding is represented by the ohmic resistance of the primary winding R_P and the primary leakage inductance L_P; the secondary winding by the ohmic resistance of the secondary winding R_S and the secondary leakage inductance L_S. The behaviour of the iron core is represented by the main inductance L_m and the ohmic losses of the core R_{Fe}. The impedance of the external burden is represented by the ohmic part R_B and the inductive part L_B. The primary values are converted to the secondary side. Here applies:

$$I'_P = I_P{}^{N_P}/{}_{N_S} = I_P/\ddot{u}, \qquad R'_P = R_P\left(N_S/N_P\right)^2 = R_P{\cdot}\ddot{u}^2, \qquad L_P = L'_P\left(N_S/N_P\right)^2 = L_P{\cdot}\ddot{u}^2$$

with the transmission ratio $\ddot{u} = N_S/N_P$

The primary winding resistance R_P and the primary leakage inductance L_P *are* not relevant for considering the accuracy of the current transformer. If the primary conductor is concentric to the core and the secondary winding is evenly distributed on the core, the secondary leakage inductance L_S can also be neglected. This reduces the equivalent circuit to the simplified circuit of Fig. 3.5.

R_{Fe} represents the effective losses in the iron core, which are caused by remagnetisation and eddy currents. L_m is the main inductance of the transformer.

The secondary current causes a voltage drop at the secondary winding (R_S and L_S) and at the external burden (R_B and L_B). This voltage U_0 leads to a magnetic flux in the iron core, which induces the magnetising current I_0. The primary current is the sum of the secondary current and the magnetising current. The magnetising current I_0 thus represents the error of the current transformer.

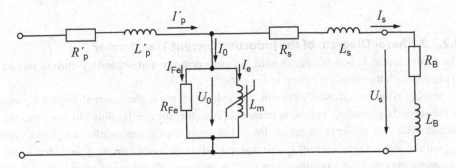

Fig. 3.4 Equivalent circuit diagram of an inductive current transformer reduced to the secondary side

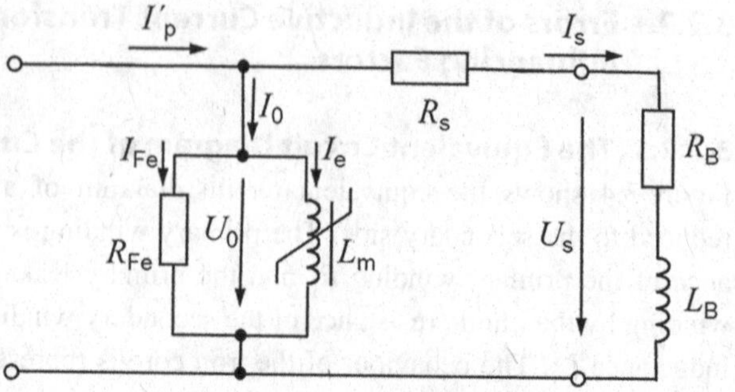

Fig. 3.5 Simplified equivalent circuit diagram of an inductive current transformer reduced to the secondary side

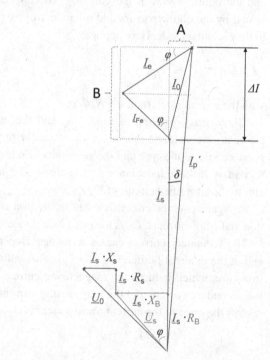

Fig. 3.6 Current transformer phaser diagram according to equivalent circuit Fig. 3.4 (referring to the secondary side)

3.2.2.2 Phasor Diagram of the Inductive Current Transformer

The error current I_0 and the phase shift between primary and secondary current can be explained with the phasor diagram in Fig. 3.6.

Starting with the secondary current, the voltage drops at the external burden U_S and at the secondary winding can be determined. With the voltage $\underline{U}_0$ thus obtained and the characteristics of the core material, the ohmic iron losses, represented by the current $\underline{I}_{Fe}$ and the magnetising current $\underline{I}_e$, are determined. If the error current $\underline{I}_0$ is added to the secondary current $\underline{I}_S$, the primary current $\underline{I}_P'$ is obtained. The diagram shows the current error ΔI and the phase shift δ between $\underline{I}_P'$ and $\underline{I}_S$.

3.2.2.3 The Error Calculation

The amplitude error of a current transformer is according to IEC 61869-2

$$\varepsilon = \frac{K_r I_s - I_p}{I_p} \cdot 100\%$$

The rated transmission ratio K_r is the ratio of the rated value of the primary current I_{pr} to the rated value of the secondary current I_{Sr}. Normally this is equal to the ratio of the secondary to the primary number of turns.

$$K_r = \frac{I_{pr}}{I_{sr}} = \frac{N_s}{N_p} = \ddot{u}$$

Calculation of the amplitude error of a current transformer:
   Based on the law of induction

$$u(t) = N_S \frac{d\Phi}{dt} = N_S \cdot A_{Fe} \frac{dB}{dt} = N_S \cdot A_{Fe} \frac{d\hat{B} \sin \omega t}{dt} = N_S \cdot A_{Fe} \cdot \omega \hat{B} \cos \omega t$$

$$\Rightarrow \quad \hat{u} = \sqrt{2} \cdot U_0 = N_S \cdot A_{Fe} \cdot \omega \cdot \hat{B}$$

the peak value of the flux density in the current transformer core is calculated as

$$\hat{B} = \frac{\sqrt{2} U_0}{A_{Fe} \omega N_s}$$

with:

A_{Fe} cross section of the iron core
ω angular frequency $2\pi f$
N_S number of turns of the secondary winding.

The equivalent circuit diagram Fig. 3.5 shows:

$$U_0 = I_s \left(\sqrt{R_B^2 + \omega L_B^2} + R_s \right)$$

and with the burden, indicated as output power $S_B = \sqrt{R_B^2 + \omega L_B^2} \cdot I_S^2$, of the current transformer in VA:

$$U_0 = \frac{S_B}{I_s} + R_s I_s$$

Thus, the magnetic flux density:

$$\hat{B} = \frac{\sqrt{2}}{A_{Fe} \omega N_s} \left(\frac{S_B}{I_s} + R_s I_s \right)$$

With the magnetising current $I_0 = I_S - I_p'$, the magnetic field in the core can be determined $H = \frac{N_S I_0}{l_{Fe}}$ from the magnetisation curve in $H = f(\hat{B})$ Fig. 3.5 (l_{Fe} is the mean iron path length). With, $H = \frac{B}{\mu}$ the magnetising current becomes:

$$I_0 = \frac{\hat{B}}{\mu} \cdot \frac{l_{Fe}}{N_S} = \frac{\sqrt{2} l_{Fe}}{A_{Fe} \omega N_S^2 \mu} \left(\frac{S_B}{I_s} + R_s I_s \right)$$

The amplitude error $\varepsilon = \frac{I_s - I_p'}{I_p'} \cdot 100\% = \frac{-I_0}{I_p'} \cdot 100\%$ can thus be calculated:

$$\varepsilon = \frac{-\sqrt{2} l_{Fe} \cdot 100\%}{I_p' A_{Fe} \omega N_S^2 \mu} \left(\frac{S_B}{I_s} + R_s I_s \right)$$

The error is always negative if no winding adjustment has been made (see Fig. 3.10 in Sect. 3.2.2.5). The error decreases with increasing core cross section A_{Fe} and increases with increasing iron path length l_{Fe}, i.e., core diameter. Furthermore, the error depends on the connected burden. In the linear range of the magnetisation curve (μ is constant) there is therefore a linear relationship between the amplitude error ε and the burden S_B. If the secondary winding is short-circuited (burden $S_B = 0$), the error does not become zero due to the losses in the secondary winding R_S.

Calculation of the phase error of a current transformer:
The phase error δ can be calculated from the phasor diagram Fig. 3.6.
   The tangent of the phase error δ results from

$$\tan \delta = \frac{A}{B + I_s} = \frac{I_e \cos \varphi - I_{Fe} \sin \varphi}{I_e \sin \varphi + I_{Fe} \cos \varphi + Is}$$

And the phase error δ to

$$\delta = arctg \left(\frac{I_e \cos \varphi - I_{Fe} \sin \varphi}{I_e \sin \varphi + I_{Fe} \cos \varphi + Is} \right)$$

The angle φ is the phase angle between I_S and U_0, which in the first approximation corresponds to the phase angle of the burden (between I_S and U_S).
   According to the IEC 61869-2 standard [34], for a burden greater than or equal to 5 VA, the power factor $\cos \varphi = 0.8$.
   For the special case of a purely ohmic burden ($\varphi = 0$), with $\cos \varphi = 1$ and $\sin \varphi = 0$, the phase error is given by:

$$\delta = arctg \left(\frac{I_e}{I_{Fe} + Is} \right)$$

The corresponding simplified phasor diagram is shown in Fig. 3.7.

Fig. 3.7 Simplified phaser diagram for pure resistive burden

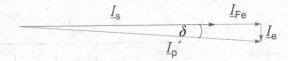

3.2.2.4 Influence of the Burden on the Current Transformer Error

Figure 3.9 shows the relationship between the current transformer error ε and the burden S_B. When the current transformer is saturated, the permeability of the core μ is strongly reduced (see Fig. 3.8) and thus, the magnitude of the error increases significantly.

For the example in Fig. 3.9 the error was calculated for a core of silicon iron with a cross-section $A_{Fe} = 4$ cm^2 and an iron path length of 0.628 m (diameter 0.2 m), a primary current $I_p = 1000$ A, a secondary current $I_S = 1$ A and a secondary number of turns of 1000.

3.2.2.5 Winding Adjustment of the Current Transformer

By adjusting the secondary winding number N_S, the secondary current can be deliberately changed, and the amplitude error can be shifted to positive values.

If, for example, in the case of a current transformer with a rated transformation ratio of $I_p/I_S = 500/1$, the winding ratio is adjusted to $N_S/N_p = 499/1$, the secondary current increases by $1/500 = 0.2\%$, i.e., the transformer error is shifted by $+0.2\%$. Now the error will increase up to $+0.2\%$ with negligible burden (Fig. 3.10).

This adjustment of the number of windings is used to make better use of the range of permissible error according to the standard.

IEC 61869-2 requires that the error limits are not exceeded in a range from ¼ of the rated burden to the rated burden. If the error at rated burden is outside the permitted accuracy limits (Fig. 3.10a), the error pattern can be shifted to the positive by adjusting the number of secondary windings, thus fulfilling the requirement of the standard (Fig. 3.10b).

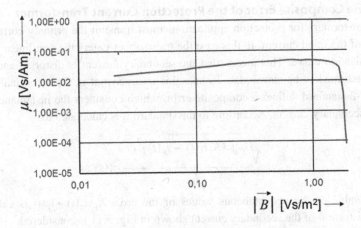

Fig. 3.8 Permeability $\mu = B/H$ of an iron core as a function of magnetic flux density B

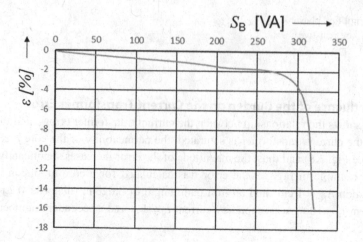

Fig. 3.9 Dependence of the amplitude error on the connected burden

If, however, the adjustment of the secondary winding number shifts the error into positive to such an extent that, if the burden is neglected, the error is greater than permitted, the condition that the error in the burden range between ¼ and the full rated burden is within the limits may be met, but if the transformer is underloaded with less than ¼ of the rated burden, the error falls out of accuracy class. Therefore, in the case of a current transformer core with winding compensation, care must be taken to ensure that the connected burden is not less than ¼ of the rated burden. In contrast to a current transformer core without winding adjustment, where a smaller burden leads to a smaller error and a low burden is not a problem in terms of accuracy. However, for all current transformers, care must be taken to ensure that the overcurrent limiting factor FS becomes greater with a smaller burden (see Sect. 3.2.3).

3.2.2.6 The Composite Error of the Protection Current Transformer

Current transformers for protection applications must transmit the primary current up to a multiple of the rated current. In this case the transformer enters the non-linear range of the magnetisation curve. This means that the secondary current is distorted and can no longer be assumed to be sinusoidal. To take this into account in the error analysis, the IEC 61869-2 standard defines a composite error, which considers the instantaneous values of the secondary current. According to the standard it is calculated to:

$$\varepsilon_c = \frac{\sqrt{\frac{1}{T} \int \left(K_r i_s(t) - i_p(t) \right)^2 dt}}{I_p} \cdot 100\%$$

So, a rms value of the instantaneous values of the error $K_r \cdot i_S(t) - i_p(t)$ is calculated. Thus, the distortion of the secondary current shown in Fig. 3.11 is considered.

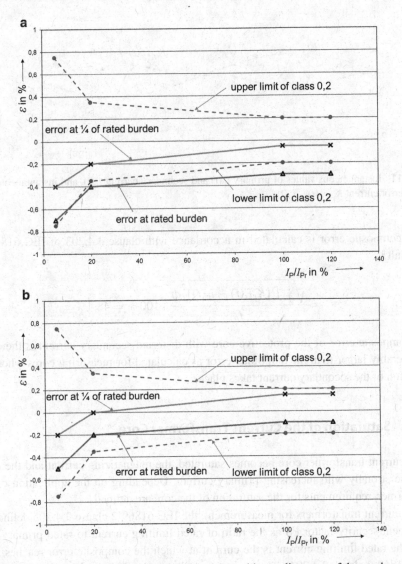

Fig. 3.10 Error of a current transformer 500 A to 1 A **a** without adjustment of the number of turns **b** with adjustment of the number of turns to 499:1

If in the example in Fig. 3.11 the ratio error is calculated from the rms values of the primary and secondary current (I_p and I_S) according to clause 3.4.3 of IEC 61869-2, the result is:

$$\varepsilon = \frac{K_r I_s - I_p}{I_p} \times 100\% = 1{,}7\%$$

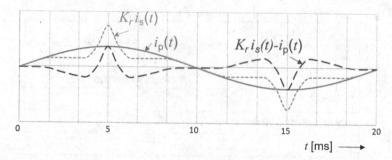

Fig. 3.11 Instantaneous values of primary $i_p(t)$ and secondary current $i_s(t)$, and instantaneous values of error current $K_r \cdot i_s(t) - i_p(t)$

If the composite error is calculated in accordance with clause 3.4.203 of IEC 61869-2, the result is:

$$\varepsilon_c = \frac{\sqrt{\frac{1}{T} \int \left(K_r i_s(t) - i_p(t) \right)^2 dt}}{I_p} \cdot 100\% = 48{,}4\%$$

The composite error of the protective core with distorted secondary current is therefore considerably larger than the amplitude error as calculated for measuring cores where no distortion of the secondary current takes place.

3.2.3 Saturation of the Current Transformer Core

If the current transformer core becomes saturated, the magnetizing current and the error increase strongly with increasing primary current. Depending on the application of the transformer, requirements for the saturation of the core are defined.

For current transformers for measurement, the IEC 61869-2 clause 3.4.205 defines an instrument security factor FS as the ratio of rated limiting current to rated primary current. The rated limiting current is the current at which the composite error reaches 10% (IEC 61869-2 clause 3.4.204). The current transformer should be saturated if the primary current exceeds a value of FS times the rated current. In this case it is avoided that the secondary current becomes too high when the primary current is high (e.g., in case of a short circuit). This limitation protects the connected secondary devices against high input currents (Fig. 3.13a).

When using such current transformers, the connected burden must be considered. If a lower burden than the rated burden is connected (R_B greater than rated value), the instrument security factor FS (instrument security factor) will be higher than intended (Fig. 3.12). This means that the current transformer goes later into saturation and in the event of a short circuit higher currents are applied to the input of the measuring devices.

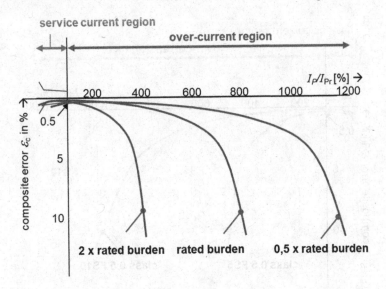

Fig. 3.12   Burden influence on CT saturation

For current transformers for protection purposes the saturation should not start before a defined value of the primary current. IEC 61869-2 clause 3.4.208 defines an accuracy limit factor (ALF) as the ratio of the rated accuracy limit current to the rated primary current. The rated accuracy limit current is the current up to which the requirement for the composite error is fulfilled (IEC 61869-2, clause 3.4.207). This definition ensures that the high current can be measured in the event of a short circuit and that the connected protective relays reliably detect the short circuit (Fig. 3.13b). Like the FS, the ALF also depends on the connected burden. For smaller burden, the ALF is increased (Fig. 3.12).

Figure 3.12 shows the influence of the connected burden. With a lower burden the current transformer core is saturated at higher primary currents, which increases the instrument security factor FS or the accuracy limiting factor ALF.

Figure 3.13 shows the error curves of measuring and protective cores as a function of the primary current with the limits according to IEC 61869-2 for an FS or ALF of 5 and 10.

3.2.4   Influence of the Leakage Flux in the Current Transformer Core

In an ideal current transformer with short-circuited secondary winding without internal losses (winding resistance $R_S = 0$ and leakage inductance $L_S = 0$) and without external burden, the voltage U_0 and thus the magnetising current I_0 *is* zero (see equivalent circuit

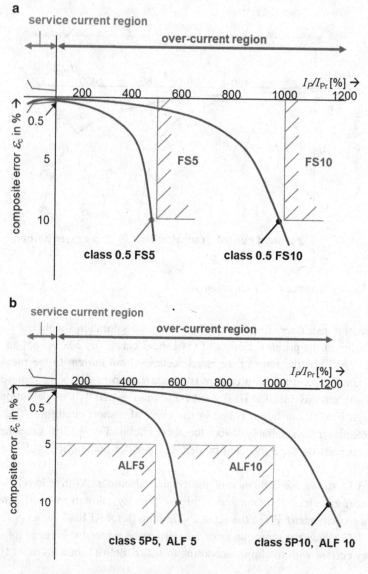

Fig. 3.13 Saturation requirements for **a** Measuring cores (overcurrent limiting factor FS) **b** Protection cores 5P (accuracy limiting factor ALF)

diagram 3.5). With a primary conductor in the centre of the toroidal core and a secondary winding evenly distributed around the circumference, the primary current in the core will generate a magnetic flux Φ_P of the same magnitude throughout the core, which is the same magnitude as the magnetic flux Φ_S, generated by the secondary current but in the opposite direction (Fig. 3.14a). The total flux as the sum of these two will therefore be zero.

Fig. 3.14 Magnetic flux in
a current transformer core
a ideal current transformer
without consideration of
winding resistance **b** real
current transformer with
winding resistance and burden

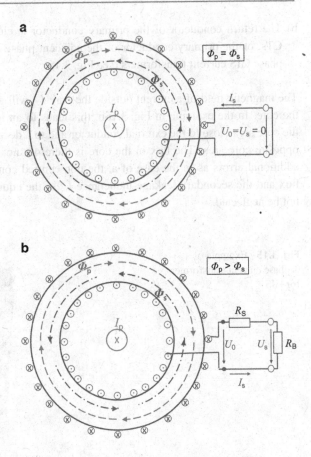

However, in the real current transformer, there is a voltage drop U_0 (see Fig. 3.5) caused by the resistance of the secondary winding R_S and the external burden R_B (Fig. 3.14b). To generate the voltage drop U_0, a magnetic flux is required, which is generated by the magnetising current I_0. The secondary current is reduced by this error current and therefore the sum of the primary and secondary flux is no longer zero.

In some applications the magnetic flux is not constant along the circumference of the core, which can lead to additional errors of the current transformer.

a) Primary conductor is not centred in the toroidal core:

Only if the primary conductor is placed exactly in the centre of the toroidal core, the primary magnetic flux is constant throughout the core. If this is not the case, the voltage induced in the secondary winding will not be evenly distributed around the core, leading to additional errors (see Fig. 3.16a). This additional error creates an additional flux, the secondary leakage flux. This is represented by the secondary leakage inductance L_S in the equivalent circuit in Fig. 3.4 and cannot be neglected in this case.

b) The return conductor of the primary conductor is close to the core as in dead tank CTs, or the primary conductor of the adjacent phase is close to the core as in three-phase GIS current transformers (see Fig. 3.15):

The magnetic field of a current outside the core I_{ex} will generate a magnetic flux Φ_{Iex} in the core. In the example in Fig. 3.16b, this leads to an increase in the magnetic flux in the core area close to the external conductor and to a decrease in the magnetic flux in the opposite core area. The flux in the core is therefore not constant, which again leads to additional errors as in the case of a. the non-centred conductor, The secondary leakage flux and the secondary leakage inductance L_S in the equivalent circuit 3.4 can therefore not be neglected.

Fig. 3.15 Example of a
3-phase current transformer
for GIS

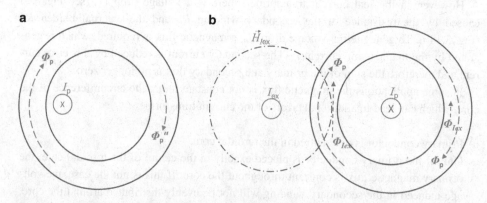

Fig. 3.16 Influence of a **a** non-centred primary conductor, $\Phi'_p > \Phi''p$ **b** external return conductor, weakens the flux on the right side and increases the flux on the left side

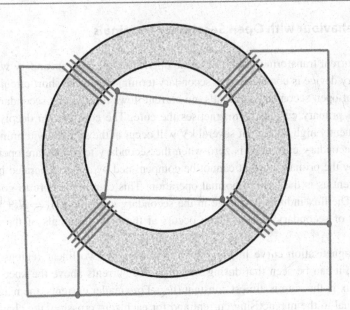

Fig. 3.17 Compensating windings

If the secondary winding is not evenly distributed around the core, the same effect
will occur. The voltage induced in the secondary winding, and therefore the secondary flux, will not be evenly distributed in the core, resulting in additional error currents.
These additional error currents will generate an additional flux, the secondary leakage
flux. This can be used to compensate for the influence of a non-centred primary conductor or an external return conductor. If it is known in which part of the core the primary
magnetic flux is increased, the resulting additional error can be compensated by applying
more turns of the secondary winding at this part of the core and correspondingly fewer
turns at the opposite part,

To compensate the influence of the above-described cases a. and b., it is possible to
apply special compensation windings to the core. Several identical windings are placed
around the core as shown in Fig. 3.17. The beginning and end of these windings are connected together. If the flux around the core is the same, the same voltage will be induced
in all the additional windings and no current will flow through them.

If the flux in the core is not constant, a voltage corresponding to the respective flux
at this point is induced in the individual compensation windings, which is not the same.
The voltage difference will drive a current through the windings whose flux in the core
compensates for the original unevenness and thus for the secondary winding a constant
flux in the core is created again and thus the additional error described above does not
occur.

These difficulties do not occur in the case of life tank CTs, as here a symmetrical primary and secondary winding in the design of the current transformer is possible.

3.2.5 Behaviour with Open Secondary Terminals

Inductive current transformers shall never be operated with open secondary windings. If no secondary device is connected, the secondary terminals must be short-circuited.

In case of open secondary terminals, no current flows through the secondary winding and the full primary current I'_p magnetises the core. The core is then deeply saturated and consequently high voltage of several kV will occur at the secondary terminals.

As the secondary current I_S is zero when the secondary terminals are open, the flux generated by the primary current cannot be compensated by a flux generated by the secondary current as is the case in normal operation. This causes the core to go deep into saturation. The flux induces a voltage in the secondary winding $u(t)_S = N_S \, {}^{d\Phi}\!/_{dt}$ (N_S is the number of secondary turns), which occurs at the open terminals of the secondary winding.

If the magnetisation curve in Fig. 3.3 is simplified by two linear regions as shown in Fig. 3.18, it can be seen that during saturation for currents above the knee point, the magnetic flux in the core is almost constant (Φ_S). For smaller currents, the magnetic flux is proportional to the magnetising current and for each zero crossing, the change in flux over time reaches a large value, resulting in a high peak voltage at the secondary terminals. This can lead to damage to the insulation.

Figure 3.19 shows the primary current $i'_p(t)$, the resulting magnetic flux in the current transformer core $\Phi(t)$ and the voltage $u_S(t)$ occurring at the open secondary terminals as a function of time. The voltage peaks occur at the zero crossings of the current when $d\Phi/dt$ has its maximum.

The voltage at the open secondary terminals depends on the primary current and the size of the magnetic core. The voltage increases with increasing core cross section and with increasing primary current. The voltage decreases with increasing core length (increasing diameter of the core). For a power frequency of 60 Hz the voltage will be 20% higher than for the same core and primary current at 50 Hz. The steepness of the magnetisation curve in the non-saturated range (i.e., the initial permeability of the core

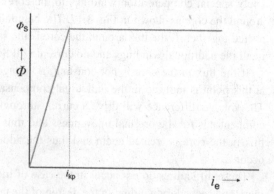

Fig. 3.18 Simplified magnetisation curve of a typical current transformer

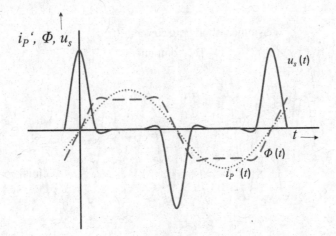

Fig. 3.19 Primary current $i'_p(t)$, magnetic flux $\Phi(t)$ and the voltage at the secondary terminals $u_S(t)$ for a current transformer with open secondary winding

μ_{ra}) is proportional to the voltage at the secondary terminals. This means that for NiFe cores ($\mu_{ra} \approx 50,000$), which are used for measuring purposes, the voltage will be about 5 times higher than for SiFe cores ($\mu_{ra} \approx 10,000$) with the same dimension [40].

In addition to the high voltage $u_S(t)$ at the secondary terminals, which can damage the insulation, the high magnetic flux $\Phi(t)$ in the current transformer core leads to large losses and possibly to overheating of the core. An overheated core will permanently change its magnetic properties and thus the accuracy of the core will be permanently changed.

3.2.6   Transient Characteristics of the Conventional Current Transformer

The short circuit current $i_k(t)$ is the sum of a sinusoidal short circuit current and a decreasing direct current. The initial value of the DC component depends on the phase angle of the short-circuit inception and has its maximum when the short-circuit is starting in the maximum of the primary current. The DC component decreases with the primary time constant T_p, which depends on the short-circuit impedance of the system ($T_p = L_p/R_p$). Figure 3.20 shows the short-circuit current curve for the maximum direct current component and a primary time constant T_p of 60 ms. Typical values of T_p are between 60 and 100 ms.

$$i_k(t) = \sqrt{2}I_{th}\left[e^{-t/T_p} + \sin \omega t\right]$$

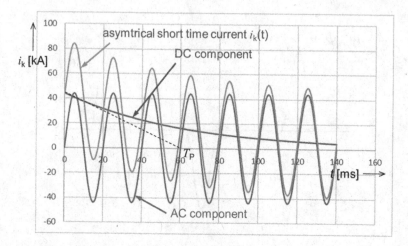

Fig. 3.20 Short-circuit current with maximum DC component

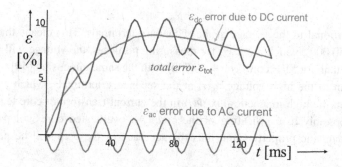

Fig. 3.21 Instantaneous value of current transformer error during short circuit

If the short-circuit current shown in Fig. 3.20 flows through the primary winding of a conventional current transformer, the instantaneous value of the error is composed of an AC component and a DC component, as shown in Fig. 3.21.

According to the standard IEC 61869-2, for the TPX and TPY class, the transient error of the sum of the AC and DC components is essential, while for the TPZ class, only the compliance of the AC component is required.

In addition to compliance with the error limits of 10% according to the standard, a remanence factor of less than 10% must be maintained for TPY and TPZ current transformers. The remanence factor K_R is the ratio between the remanence flux Ψ_r and the saturation flux Ψ_s in the current transformer core ($K_R = \Psi_r / \Psi_s$) as shown in Fig. 3.22. TPZ cores have a large air gap, therefore the remanence factor of TPZ cores is significantly smaller than 10%.

Fig. 3.22 Remanence flux ψ_r and saturation flux ψ_s of a current transformer core [34]

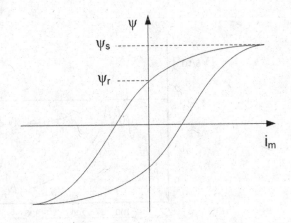

	Core type	Remanenz
I	Closed iron core, z.B. P, TPS, TPX	Up to 90 %
II	With small air gap, z.B. TPY	< 10 %
III	With large air gap, z.B. TPZ	negligible

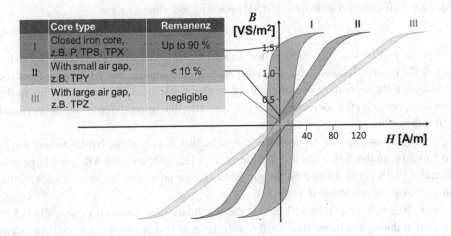

Fig. 3.23 Hysteresis curves of cores with and without air gap

To reduce the remanence factor, air gaps are built into the core. The air gap linearises the magnetisation characteristic. Figure 3.23 shows the difference in the hysteresis curve between a core without air gap (class P and TPX), with a small air gap (class PR and TPY) and with a large air gap (TPZ).

The introduction of an air gap linearises the magnetisation characteristic of the core. The main inductance L_m is reduced with the consequence that the magnetising current I_e is increased. This increases the ratio error and the error angle of the current transformer.

If, as described in patent EP 1 851 556 B1 [41], a capacitor is connected to the secondary terminals of the current transformer, the phase displacement caused by the air gap can be compensated.

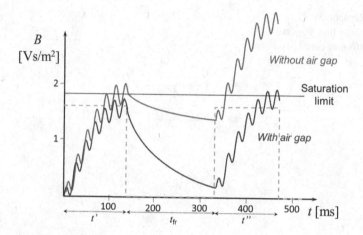

Fig. 3.24 Flux density B during a C–O–C–O cycle with and without air gap

The air gap of the core reduces the secondary time constant $T_S = L_S/R_S$ (L_S is the sum of the main inductance and secondary leakage inductance, and R_S is the sum of the resistance of the secondary winding and the burden resistance). The magnetic flux in the core decreases with the secondary time constant after the interruption of the short circuit by the circuit breaker.

Without an air gap, the reduction of magnetic flux is very slow, typical values for T_S are 5 to 20 s, so the flux starts at a high level after reconnection and will quickly go into saturation. TPX cores without an air gap are therefore only used for specifications without automatic reconnection (C–O).

With the small air gap in TPY cores, the secondary time constant is typically 0.5 to 1 s, so that during the pause time until restart the core is demagnetised to such an extent that after reconnecting the core is not saturated within the time to the accuracy limit t''_{al}. The magnetic flux density B as a function of time for an automatic reconnection cycle (C–O–C–O) is shown in Fig. 3.24. If the flux density rises above the saturation limit, the error is greatly increased.

For TPZ cores the air gap is large, the secondary time constant is about 60 ms. During the pause time t_{fr} before reconnecting, the flux density is thus reduced to almost zero, and magnetisation starts from the beginning at the second short circuit. The requirement for the phase displacement of a TPZ core of 180 min corresponds to a secondary time constant of 60 ms for a system frequency of 50 Hz and 50 ms for a system frequency of 60 Hz.

In the current transformer standard IEC 61869-2, the rms value of the AC component of the short-circuit current is defined as I_{psc}. Up to this current, the limits of accuracy must be met. There is also the definition of the thermal short-circuit current I_{th}, which must be withstood for a certain time (typically 1 s or 3 s). It describes the thermal load

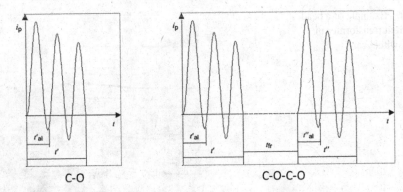

Fig. 3.25 Short-circuit duty cycles without (C–O) and with (C–O–C–O) automatic reconnection [34]

during the short circuit. The short circuit current I_{psc}, which is responsible for the accuracy, is smaller than the thermal short circuit current I_{th} ($I_{psc} \leq I_{th}$).

The dynamic current I_{dyn} is defined as the peak value of the primary current that the current transformer must withstand. This is related to the mechanical load on the current transformer during the short circuit.

The mechanical load during the dynamic short-circuit current I_{dyn} must be considered, especially in the case of current transformers with several primary windings and with reconnectable primary windings.

For current transformers of class TPX, TPY and TPZ, limits of transient errors are required during a defined short-circuit duty cycles. Simple short-circuit duty cycles C–O or short-circuit duty cycles with automatic reconnection C–O–C–0 are defined. The times t' and t'' are the duration of the first and second short circuit respectively, the times t'_{al} and t''_{al} are the specified times in which the accuracy requirements must be met. The time t_{fr} is the pause time between the two short-circuits during an automatic reconnection (see Fig. 3.25).

3.2.7   Design Types of Conventional Current Transformers

3.2.7.1 Head Type Current Transformer (Life Tank Design)

Most modern conventional current transformers are designed as head type current transformers (see Fig. 3.26). In these, the transformer cores with the secondary windings are housed in a metal head housing at high voltage potential (life tank design, see Fig. 3.27). In most cases the primary conductor is realised as an aluminium bar which is passed through the cores. Designs with several primary windings or designs with primary side reconnectable transmission ratio are described below.

Fig. 3.26 Example of a head
type current transformer in
420 kV switchgear

Fig. 3.27 Sectional view of a
head type current transformer

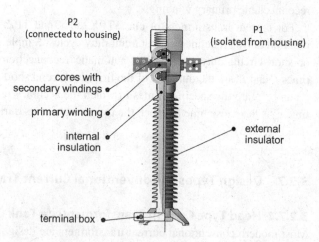

P2
(connected to housing)

P1
(isolated from housing)

cores with
secondary windings

primary winding

internal
insulation

external
insulator

terminal box

The transformer cores are installed in a core housing, which is connected to the bushing tube. The metallic core housing serves as the earth-side electrode of the insulation. The leads of the secondary windings are led through the bushing tube to the terminal box at the base of the transformer. The bushing tube connects the core housing with the earth potential.

The core housing must be insulated against the primary conductor, which is at high voltage, and against the head housing. Today's common insulation materials are oil-paper or SF_6.

The primary conductor is connected to terminals P1 and P2. Terminal P2 is directly connected to the head housing, whereas terminal P1 is insulated from the head housing. This insulation must be dimensioned for the voltage drop in the event of a short-circuit current and other over-currents. In the case of a head type current transformer with one primary winding, i.e., a straight short bar between the terminals, the ohmic resistance and the primary inductance are small and therefore the voltage drop is low (low reactance type) [42].

Because of the short straight primary conductor of the head type current transformer, the ohmic resistance R_P and the primary leakage inductance L_P can be neglected. Forces on the primary conductor by the short-circuit current and the heating of the transformer by the primary current are therefore very small in the head type current transformer.

Both the dielectric strength and the forces occurring must be considered when designing a head type current transformer with several primary windings.

Due to the symmetrical design of the head type current transformer with the primary conductor in the centre of the transformer cores, the secondary leakage flux in the current transformer core can be neglected, and thus high accuracy can be achieved (see Sect. 3.2.4).

To use the current transformer flexibly or to be able to use the same current transformer when the primary current changes later, a primary reconnection is required for some transformers. Usually 1:2 or 1:2:4 reconnections are used.

1:2 reconnection:
In the 1:2 reconnection, as shown in Figs. 3.28 and 3.29, the primary conductor is realised by a conductor tube (2) instead of a bar, in which an additional conductor bar (1) is inserted. The conductor tube is connected to the head housing on the P2 side and directly connected to P1 insulated from the head housing on the P1 side. The connection P2 is connected to the inner conductor bar. This bar can be moved and on the P1 side it can be connected either to terminal P1, or to the head housing.

Fig. 3.28 Principle of 1:2 reconnection

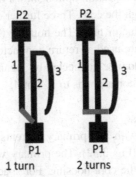

1 inner contact bar
2 conductor tube
3 housing as return conductor

1 turn   2 turns

Fig. 3.29   Primary reconnection 1:2 on head type current transformer

For one primary turn, the bar (1) is connected directly to connection P1. The current thus flows once through the core in the inner conductor bolt (1) from connection P1 to connection P2.

For two primary turns the bar (1) is connected to the housing (3). The current thus flows from connection P1 in the conductor tube (2) through the core, then flows back via the housing (3) outside the core and through the inner conductor (1) a second time through the core and to connection P2. The primary number of turns N_P is thus two and the transformation ratio N_S/N_P is changed to half the value.

1:2:4 reconnection:

For the 1:2:4 primary-reconnection option shown in Figs. 3.30 and 3.31 a conductor tube (2) is again passed through the core. Three further conductors (1) are insulated from each other placed in this conductor tube. The housing (3) and two external return conductors (4) outside the housing are used as return conductors. The three primary conductors and the conductor tube can now be connected by means of connecting pieces (5) so that the primary number of turns corresponds to 1, 2 or 4.

Many primary windings:

For special applications with low primary currents, current transformers with many primary windings are used (Fig. 3.32). The primary winding is wound in the head housing around the insulation of the core housing. For space reasons, however, the cross-section

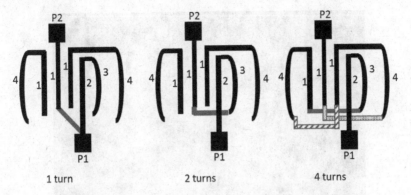

1 turn   2 turns   4 turns

Fig. 3.30 1:2:4 reconnection principle

Fig. 3.31 Primary reconnection 1:2:4 on the head CT (Pos 1: Inner primary conductor; Pos 2: Primary conductor tube; Pos 3: Housing as return conductor; Pos 4: External return conductor; Pos 5: Changeover lugs)

of the primary winding becomes small and therefore the use of such current transformers is limited regarding the short-circuit current occurring in the system. In modern head type current transformers, up to 50 or more primary windings can be realised.

3.2.7.2 The Hairpin and the Eye-Bolt Current Transformer (Dead Tank Design)

In contrary to the head type current transformer, the cores with the secondary windings of the hairpin current transformer and the eye-bolt current transformer are in a housing at

Fig. 3.32 Head CT with several primary windings

ground potential. For this purpose, the primary conductor must be led through the insulator into the housing and back again. For this reason, the primary conductor for high-voltage transformers is several metres long. In these designs the insulation is applied to the primary conductor. Figure 3.33 shows the basic structure of these current transformer types. In the hairpin current transformer, the primary conductor is U-shaped (see Fig. 3.33a), in the eye-bolt transformer the forward and return conductor of the primary conductor is insulated together (see Fig. 3.33b)

An advantage of these designs is the low centre of gravity, which is advantageous for seismic requirements. However, this requires an insulator with a larger internal diameter to accommodate the primary conductors, which results in a higher weight and larger oil volume of the current transformer.

A further advantage is seen in a possible machined insulation of the primary conductor, which is designed like a cable in the hairpin transformer.

However, the long primary conductor and the design of these types has several disadvantages compared to the head type current transformer:

a) The losses of the primary conductor (R_p) and the primary leakage inductance (L_p) can no longer be neglected (high reactance type [42]). There is significantly higher ohmic heating ($I_p^2 R_p$) of the current transformer during operation and in the event of a short circuit. Special measures are required to cool the transformer, for example by means of cooling fins in the head (see Fig. 3.34).
Using the example of a current transformer with a primary current of 3600 A, the losses in a head type current transformer and a hairpin current transformer are calculated below:

Fig. 3.33 Basic structure of the **a** hairpin current transformers and **b** the Eye-Bolt current transformer

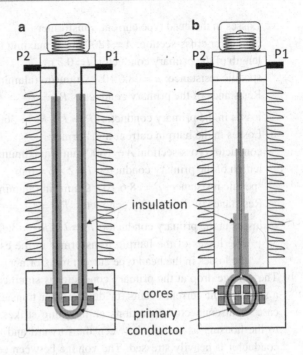

Fig. 3.34 Cooling of a hairpin transformer by cooling fins (ABB current transformer 362 kV, 3600 A in Malaysia)

i. Losses in the head type current transformer:

conductor cross-section: $A = 1800$ mm^2 (maximum current density 2 A/mm^2)

length of the primary conductor: $l = 0.5$ m

specific resistance: $\rho = 28.6 \ 10^{-3}$ Ωmm^2/m (aluminium)

Resistance of the primary conductor: $R_P = \frac{\rho \cdot l}{A} = \frac{28.6 \cdot 10^{-3} \cdot 0.5}{1800} = 7.94 \cdot 10^{-6} \Omega$

losses in the primary conductor: $P = I^2 \cdot R_P = 3600^2 \cdot 7.94 \cdot 10^{-6} = 102.9$ W

ii. Losses in the hairpin current transformer:

conductor cross-section: $A = 1800$ mm^2 (maximum current density 2 A/mm^2)

length of the primary conductor: $l = 2 \times 3 = 6$ m

specific resistance: $\rho = 28.6 \ 10^{-3}$ Ωmm^2/m (aluminium)

Resistance of the primary conductor: $R_P = \frac{\rho \cdot l}{A} = \frac{28.6 \cdot 10^{-3} \cdot 6}{1800} = 95.28 \cdot 10^{-6} \ \Omega$

losses in the primary conductor: $P = I^2 \cdot R_P = 3600^2 \cdot 95.28 \cdot 10^{-6} = 1234.8$ W

The losses of the hairpin transformer in the example are 12 times higher than the losses in the head type current transformer.

b) The voltage drop at the primary conductor is significantly increased, especially when a short-circuit current occurs, or during other transient processes, for example in the case of disconnector operations or lightning strikes. The insulation on the P1 side to the housing as well as between the forward and return conductor of the primary conductor is heavily stressed. The voltage between connection P1 and the housing is often limited by special measures such as the use of surge arresters or spark gaps.

c) In the event of a short circuit, high forces occur between the forward and return conductors of the primary conductor, which must be absorbed by the design.

This force is calculated as follows:

$$F = \mu_r \mu_0 \frac{I_{dyn} \cdot I_{dyn}}{2\pi} \cdot \frac{l}{a}$$

For the example of a hairpin current transformer with a length of the forward and return conductors of 3 m each and a distance of 0,1 m between the two conductors, the force resulting from the peak of the fully displaced short-circuit current is:

$$F = 4\pi 10^{-7} \frac{Vs}{Am} \frac{100^2 \cdot 10^6 A}{2\pi} \cdot \frac{3\,m}{0.1\,m} = 60000\,N$$

where:

thermal short-circuit current: $I_{th} = 40$ kA (rms value)

dynamic short-circuit current: $I_{dyn} = 2.5 \ I_{th} = 100$ kA (peak value)

Permeability between the conductors: $\mu_r = 1$, $\mu_0 = 4\pi 10^{-7}$ Vs/Am

d) The arrangement of the primary conductor in hairpin and eye-bolt current transformers influences the accuracy of the current transformer cores. The primary current flows in the primary conductor through the transformer core and then back close to the core. This results in leakage flux in the core as described in Sect. 3.2.4, which leads to additional errors. High accuracies are therefore very difficult or even impossible to achieve with these designs. This effect can be reduced by compensating windings or unevenly wound secondary windings (see Sect. 3.2.4).

3.2.8 Dielectric Stress and Insulation of the Current Transformer

3.2.8.1 Voltage Stresses of Current Transformers

a: Load with AC Voltage

An instrument transformer in the power network is exposed to a continuous AC voltage with a power frequency of 50 Hz or 60 Hz. In the event of disturbances in the network, this voltage can also be increased for a short time. Thus, in unearthed networks, in the event of a single-phase short circuit, the voltage on the unaffected phases can rise 1.9 times and remain for several hours until the short circuit is resolved.

To check the quality of the insulation, each transformer is subjected to a routine AC voltage test. A voltage defined in the standards is applied to the transformer for 1 min (see Table 3.1). The test is passed if there are no internal breakdowns or external flashovers.

Partial discharges in the insulation can damage it severely and must therefore be avoided. According to the standard IEC 61869-1, the permitted limits are 5 PC at 1.2 times the rated voltage and 10 PC at the maximum operating voltage U_m. However, since partial discharges during operation can decompose the oil and thus damage the transformer over time, in a good insulation partial discharges up to the applied test AC voltage should not occur.

Table 3.1 Test voltages according to IEC 61869-1 (highest values in each case)

U_m in kV	1 min AC test voltage in kV rms	lightning impulse voltage 1.2/50 μs in kV peak	switching impulse voltage 250/2500 μs in kV peak
72,5	140	325	–
123	230	550	–
145	275	650	–
170	325	750	–
245	460	1050	–
300	460	1050	850
362	510	1175	950
420	630	1425	1050
550	680	1550	1175
800	975	2100	1550

The insulation of the current transformer is designed to withstand the continuous AC voltage in operation for at least 40 years without damage. The maximum voltage that an oil-paper insulation can be exposed to without partial discharges or breakdowns depends on the time during which the load is applied. The service lifetime of the oil-paper insulation is described in Sect. 1.3.6.

b: Load with Impulse Voltages

Lightning strikes and switching operations in the network can cause surge voltages applied to the current transformer. According to the IEC standard, the voltages which a transformer must withstand without damage are divided into lightning impulse voltage and switching impulse voltage (see Table 3.1). With the lightning impulse voltage with a waveform of 1.2/50 µs mainly the internal insulation is tested.

For external insulation, the slower switching impulse voltage with a waveform of 250/2500 µs becomes increasingly relevant at higher operating voltages. Testing with switching impulse voltage is required for instrument transformers with a maximum operating voltage of 300 kV (phase-phase) or higher. The switching impulse voltage must be tested under rain. The test voltage is corrected with factors to consider the atmospheric conditions pressure, temperature, and absolute humidity according to IEC 60060-1 [43].

However, faster transient voltages can also stress the instrument transformer. With disconnector operations, fast pulses with a rise time of less than 1 µs are generated during restrikes, rise times of a few ns can occur with vacuum switches as well as switching operations in GIS. These fast processes usually do not directly damage the insulation, but if the internal electrical connections are not optimal, partial discharges can occur, which decompose the oil and lead to gas formation in the oil. This leads to a general weakening of the insulation. To test the behaviour under fast transient voltages, IEC 61869-1 introduced a test with 600 chopped lightning impulse voltages. After this exposure, the increase in the decomposition gases hydrogen, methane and acetylene must not exceed the limits defined in the IEC 61869-1 standard.

c: Dimensioning of Insulation in Current Transformers

To consider the various loads on the instrument transformer, a dimensioning voltage is defined for dimensioning the insulation.

The dimensioning voltage depends on the values of the dielectric load of the instrument transformer. The AC test voltage as well as the impulse test voltage and the maximum operating voltage must be considered.

Damage to the insulation by partial discharges or breakdown occurs at significantly higher peak values when a lightning impulse voltage is applied than when an AC voltage is applied. This is due to the lower energy which is fed into the insulation during the rapid process of the impulse. Based on experience and tests with insulation arrangements, a factor of 2.3 was determined for this compared to the 1 min AC test voltage.

If an AC voltage is applied over a longer period, the insulation is damaged even at voltages lower than the AC test voltage. This is shown in the lifetime curve of the insulation (see Fig. 1.15 in Sect. 1.3.6). For a desired lifetime of more than 40 years a factor between the maximum operating voltage and the AC test voltage of 2.6 was determined.

The dimensioning voltage for the insulation structure thus results to the highest value of the following voltages:

- 1 min AC Test voltage
- Lightning impulse voltage/2.3
- Maximum continuous phase-to-ground voltage $\times$ 2.6 ($2.6 \times U_m/\sqrt{3}$)

Table 3.1 shows that for instrument transformers for higher voltages (>500 kV) the operating voltage becomes the dimensioning factor. For 550 kV instrument transformers, for example, the dimensioning is based on the maximum continuous phase-to-ground voltage of $2.6 \times 550/\sqrt{3} = 826$ kV. This value is higher than the 1 min AC test voltage of 680 kV and must therefore be used as dimensioning voltage. If this is not considered, the instrument transformer will pass the required tests without any problems, but the lifetime of the transformer can be drastically reduced.

3.2.8.2 The Insulation Design of a Conventional Head Type Current Transformer

This chapter describes the insulation design of a conventional head type current transformer. It can be divided into three parts (see Fig. 3.35): One part is the insulation of the current transformer head (1), i.e., between the core housing at earth potential and the primary conductor and the head housing at high voltage potential. Another part is the insulation of the bushing tube (2), through which the leads of the secondary windings were led to the terminal box. In this bushing insulation a potential grading along the insulator is also realised by capacitive inserts. A third part of the insulation is the transition area (3) between the head insulation and the bushing insulation.

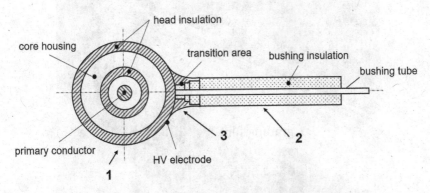

Fig. 3.35 Basic structure of insulation of conventional high voltage current transformers

For a good quality of insulation, the following boundary conditions must be observed:

- Use of high-quality insulating paper (see Sect. 1.3)
- Careful drying the insulating paper (see Sect. 3.2.8.3)
- Quality of the insulating oil (see Sect. 1.2)
- Use of degassed and dried insulating oil (see Sect. 1.2)
- Avoid oil gaps greater than 3 times the paper thickness in the insulation area 1
- Dimensioning of the occurring oil gaps in area 3 according to Fig. 3.45.

a: The Insulation of the Bushing

The insulation of the bushing and the potential grading have already been described in Sect. 2.2. The bushing insulation is wound on a semi-automatic winding machine (see Fig. 3.36). The material used is Kraft paper in the width of the bushing length. The rollers of the winding machine are heated so that moisture is removed from the paper already during winding. At the head end the insulation is carried out in steps, so that the transition to head insulation can then be realised by manual taping, as described below.

A capacitive grading is incorporated into the bushing insulation to control the potential in axial direction. This is achieved by inserting aluminium films (see Sect. 2.2). The control length corresponds to about 60% of the arcing distance of the external insulator. As shown in Fig. 3.37, the position of the grading is laid in such a way that the uncontrolled length is 25% of the arcing distance at the upper, high voltage end and 15% at the lower, earth end.

Fig. 3.36 Loading the grading layers on the semi-automatic winding machine

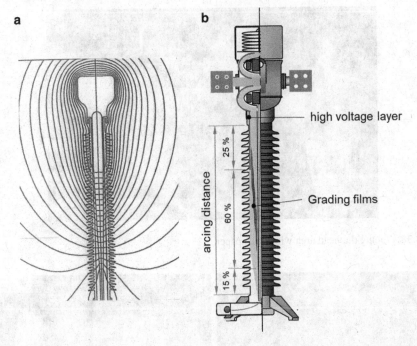

Fig. 3.37 The capacitive potential grading in the current transformer bushing **a** Potential lines, **b** Distribution of the arcing distance

b: The Insulation of the Head Area

The insulation of the head area, i.e., the insulation of the core housing is made of different materials. To achieve good insulation, mainly Kraft paper with a thickness of 100 to 150 μm is used, which is taped in strips of approx. 20 mm width around the core housing (see Fig. 3.38). To be able to apply this taping without wrinkles, the Kraft paper used must be able to stretch in the longitudinal direction. This stretchability should be 5–10%.

To compensate for any unevenness in the core housing, a layer of semi-conductive crepe paper is first applied before insulation is started.

With this taping, the given geometry of the core housing results in the insulation thickness on the inside being considerably larger than on the outside due to the greater overlap of the paper on the inside. To compensate for this, especially prepared paper inserts are attached to the outside, as shown in Fig. 3.39. These inserts are also made from Kraft paper strips. A taping of crepe paper is used to tie these inserts tightly to the insulation. Since the production of the crepe paper can lead to defects, the crepe paper portion contributes much less to the insulation than the Kraft paper portion. The proportion of Kraft paper must therefore be as high as possible. Crepe paper is only used for taping inserts.

Fig. 3.38 Taping the head area with Kraft paper

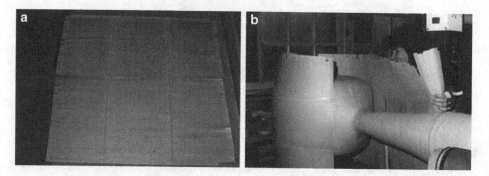

Fig. 3.39 a Paper inserts for head insulation, **b** Attaching the paper inserts

The execution of the head insulation must be even, the insulation must be well set, and care must be taken that no large oil gaps occur.

The insulation is terminated by a high voltage electrode. This is realised by firstly a layer of semiconducting crepe paper and then a metallic coating. This coating can be made of different materials:

- Metallized paper or aluminium foil, perforated to allow the oil to penetrate better into the paper underneath (see Fig. 3.40a.)
- Metallic fabric tape, (see Fig. 3.40b)
- Wire mesh (see Fig. 3.40c)

The high voltage electrode is then connected at 4 points to the housing at high voltage potential. In order not to create a short circuit winding around the core, the high voltage electrode must be interrupted at one place.

Fig. 3.40 Different types of high-voltage covering **a** perforated metallised paper **b** woven metallic tape **c** wire mesh

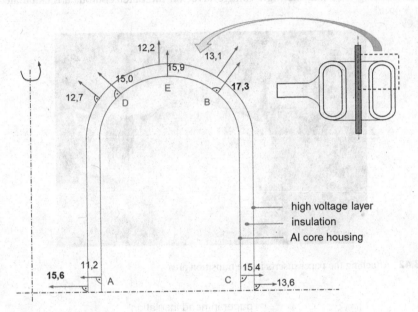

Fig. 3.41 Electric field in kV/cm in the head insulation of a current transformer

The field strength inside the insulation is determined by the geometry of the core housing and the thickness of the insulation. Figure 3.41 shows an example of a field plot of the head insulation.

The maximum field strengths occur on the inside at the high voltage electrode opposite the primary conductor (point A), and on the outer radii of the core housing (point B).

For a well-executed insulation with small oil gaps, the maximum field strength at dimensioning voltage can be up to 17.5 kV/mm. The average field strength should not exceed 13.5 kV/mm, which results in an insulation thickness of at least 17 mm for a transformer with maximum operating voltage 123 kV and AC test voltage of 230 kV.

c: The Insulation of the Transition Area

To isolate the transition area between the bushing and the head insulation, inserts made of Kraft paper strips are attached to the steps of the bushing. As shown in Fig. 3.42, these inserts extend into the head area. By alternately attaching these inserts and the insulation layers of the head area, the papers overlap, and longer oil gaps are avoided in the insulation (see Fig. 3.43).

Due to the drying process before impregnation with mineral oil, the paper inserts attached to the steps shrink and an unavoidable oil gap is formed at each step of the bushing. For the design of the steps in the bushing, the field strengths between the earthed bushing tube and the high voltage layer on the outer contour are calculated. In

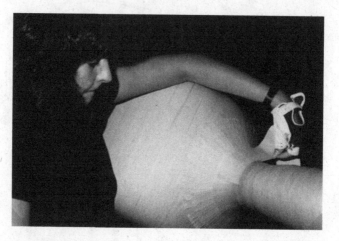

Fig. 3.42 Attaching the paper inserts in the transition area

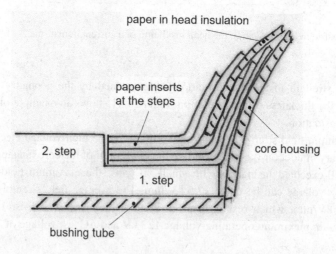

Fig. 3.43 Basic arrangement of inserts on steps and overlap with head insulation

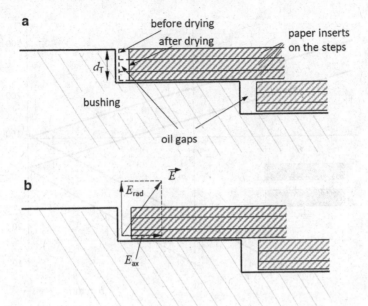

Fig. 3.44 **a** One step of the bushing in the insulation of the transition area **b** with relevant field strengths E_{rad} and E_{ax}

axial as well as radial direction these field strengths must not become too large. These field strengths are determined from the geometry of the transition area.

Figure 3.44a shows a step in the bushing with the strip inserts attached. The relevant field strengths in this area are shown in Fig. 3.44b, divided into their axial and radial components.

The number of steps, their length and depth must be selected in such a way that the oil gaps and the sliding distances at the step surface are not overloaded and cannot lead to partial discharges.

The radial field strength E_{rad} at each step determines the maximum allowed height of the step. The maximum permissible field strengths are shown in Fig. 3.45 as a function of the step height.

When calculating the radial field strength E_{rad}, it must be considered that the field strength component, which is perpendicular to an interface, changes according to the ratio of the dielectric constants. The following dielectric field constants are to be considered:

- in oil: $\varepsilon_{roil} = 2.2$
- in the bushing insulation: $\varepsilon_{roil\text{-}paper} = 3.5$
- in the hand taping of the head insulation: $\varepsilon_{rtaping} = 2.8$

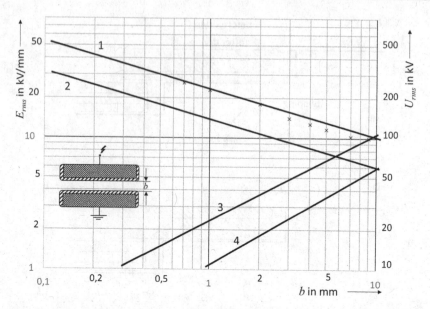

Fig. 3.45 Breakdown strength and maximum allowable field strength in an oil gap between insulated electrodes (x—measured breakdown strength)

At the transition to the bushing insulation, the field strength in the oil gap is thus increased by a factor of 3.5/2.2, at the transition to hand taped insulation, the field strength in the oil gap is increased by a factor of 2.8/2.2.

Curve 1 in Fig. 3.45 shows the rms value of the breakdown field strength in an oil gap as a function of the oil gap depth b with degassed oil and insulated electrodes determined with AC voltage at 50 Hz and a homogeneous field. The dimensioning limits of the rms value of the field strength are derived from this and shown in curve 2.

Curve 3 in Fig. 3.45 shows the breakdown voltage calculated from the breakdown field strength and curve 4 shows the limit values of the dimensioning voltage of an insulation with oil gaps as a function of the oil gap length.

In addition to partial discharges in the oil gaps, sliding discharges can also occur on the step surface of the bushing. The length of the steps must therefore be dimensioned so that the maximum permissible field strengths in the axial direction E_{ax} are not exceeded. Figure 3.46 shows the maximum tangential field strength on a paper surface as a function of the possible length of the sliding distance.

The length of each step l, as well as the total length of the possible sliding distance along all steps $a+b$ must be designed according to the curve in Fig. 3.46.

The left part of the dimensioning curve 3.46a shows the maximum average tangential field strength along one bushing step of length l (see sketch 3.45b).

For the maximum average tangential field strength along the entire steps, the right part of the dimensioning curve Fig. 3.46a applies. The maximum average tangential field

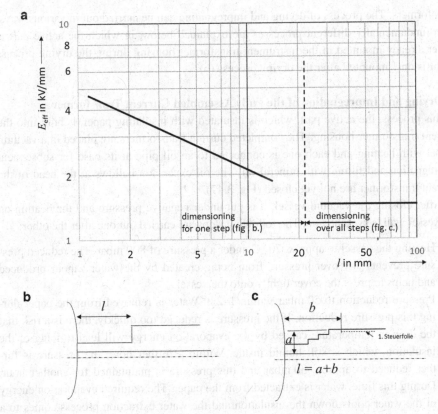

Fig. 3.46 Flashover strength and maximum permissible tangential field strength of an oil-paper surface. **a** Dimensioning curve **b** individual bushing step **c** overall bushing steps

strength along the whole steps (see Fig. 3.46c) is calculated to $\left.\vec{E}\right|_{mittel} = U_{dim}/(a+b)$. This must not exceed the value 1.14 kV/mm (see right side of dimensioning curve 3.46a).

3.2.8.3 Drying and Impregnation of the Paper Insulation

Insulating paper normally contains about 7% water by weight. This increases the losses in the finished insulation, which leads to an increased dissipation factor $\tan \delta$ and reduces the dielectric strength of the oil-paper insulation. The dissipation factor $\tan \delta$ is a measure of the losses in the insulation. It indicates the ratio of the active current I_R to the capacitive current I_c or between the losses P_V to the capacitive reactive power P_C, $\tan \delta = I_R/I_C = P_V/P_C$ [10].

To get a good quality with small $\tan \delta$ of the oil-paper insulation, the paper must first be dried. By adding heat, the water contained in the paper evaporates and is then removed from the insulation by vacuum. The dry paper insulation is then impregnated with mineral oil, which leads to the very proven good insulation of the instrument transformers. The process of drying and impregnating can be carried out in various ways. Two fundamentally different processes are explained below, in which the active part is either already mounted in the instrument transformer housing during the drying process (a) or is only mounted after the drying process (b).

a: Drying and Impregnation of the Fully Assembled Current Transformer

In this process, the active part, which is insulated with insulating paper, is built into the current transformer housing. The completed current transformers are placed in a vacuum vessel with heating and each one is connected to an oil pipe at its base for subsequent impregnation and filling with mineral oil. The oil expansion bellows at the head of the current transformer are not yet closed (Fig. 3.47).

After closing the vacuum vessel, it is put under negative pressure and the heating on the vessel wall is switched on. The following steps are carried out one after the other:

1. Heating the vessel to approx. 70 °C under a pressure of 800 mbar. The reduced pressure prevents any over pressure from being created by the water vapour produced and helps to press the cover tightly onto the vessel.
2. Pressure reduction to 50 mbar within 1–2 h. Water is removed from the paper during this pressure reduction. If the pressure is reduced too quickly, there is a risk that the drop in temperature caused by the evaporation energy will lead to icing of the insulation, which is still humid inside. Within a further hour, the pressure is further reduced to approx. 30 mbar and this pressure is maintained for another hour. During this time, water is extracted from the paper. The required evaporation energy of the water cools down the insulation and the water extraction process comes to a standstill.
3. Ventilation of the vessel to increase the temperature inside the insulation: The pressure is increased to 1 bar and maintained for 6 h. This allows the temperature of the insulating paper to be increased. Under vacuum, the heating in the boiler wall is

Fig. 3.47 a Instrument transformers mounted in vacuum vessel for drying and impregnation **b** Connection of the oil line to the transformer

not able to heat the paper in the instrument transformer sufficiently. For this reason, these intermediate ventilations are carried out to bring the temperature in the insulation to the desired temperature of 120 °C.

4. Repeat steps 2 and 3 until the temperature of 120 °C is reached at the insulation.
5. Further water removal by reducing the pressure. The pressure is reduced to ≤0.04 mbar by switching on a larger vacuum pump.
6. After reaching the vacuum of ≤0.04 mbar, the post-drying phase takes place. During this phase, evacuation continues for a time that depends on the thickness of the insulation. At the end of the post-drying time, the pressure is ≤0.01 mbar.
7. At the end of the drying process, the vessel is cooled down to 70 °C to start impregnation.
8. At a maximum pressure of 0.4 bar, the instrument transformers are now filled with mineral oil via the oil pipes connected to the base and the paper is impregnated. A maximum amount of oil per unit of time is filled into the instrument transformer so that the rate of rise of the oil level is 40 cm/hour (the rate of rise was determined by tests). This oil filling rate is calculated and set for each type of instrument transformer.
9. Once the instrument transformer is filled with oil, the post-impregnation phase begins. The impregnation process will continue to allow oil to penetrate the paper and thus lower the oil level in the instrument transformer. For 12 h, any oil that may be required is topped up.
10. Finally, the vessel is opened and at each instrument transformer the expansion bellows is filled with oil and hermetically sealed.

The temperature curve at the insulation and the pressure curve in the vessel during the drying process (steps 1 to 6) is shown in Fig. 3.48.

Once the transformers have been unloaded from the vessel, they are left for a few days (depending on the thickness of the insulation) before they are subjected to dielectric tests and partial discharge measurement.

b: Drying of the Insulated Active Part and Subsequent Assembly

In this process the finished insulated active part is hung in a vacuum vessel for drying and the paper insulation is dried in a process corresponding to steps 1–6 of the above process.

After drying, the active part is removed from the vessel and mounted in the instrument transformer housing. The instrument transformer then comes to the impregnation station (see Fig. 3.49). It is particularly important to ensure that the active parts are mounted quickly in the instrument transformer housing, as the dried active parts can very quickly reabsorb moisture from the ambient air. No more than 4 h should elapse between the time the dry active part is removed from the drying vessel and the closure of the housing. Excessive moisture absorption during assembly can be noticed during

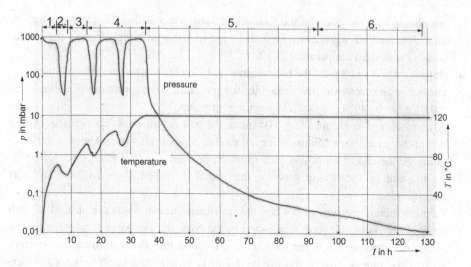

Fig. 3.48 Pressure and temperature profile during the drying process (steps 1 to 6)

Fig. 3.49 Station for filling with oil

testing by an increased dissipation factor tan δ. Experience shows a scattering of the tan δ between 2 and $4 \cdot 10^{-3}$ with this process.

At the filling station, the fully assembled current transformer is connected at the head with a vacuum hose and at the base with the oil pipe.

Then the following steps for impregnation are carried out:

1. The pressure in the instrument transformer is reduced to 0.1 mbar and held for 48 h. During this process, water is removed from the insulation, which was absorbed by the humidity during assembly.
2. Oil is filled in the instrument transformer and impregnates the insulation. The oil filling rate dependent on the size of the instrument transformer is set and the instrument transformer is completely filled.
3. Once the instrument transformer is filled, it is placed under pressure of 2.6 bar absolute to support the penetration of the oil into the paper and thus the impregnation. Small instrument transformers are kept under pressure for 24 h, large instrument transformers ($U_m > 245$ kV) for 72 h.
4. Finally, the expansion bellows is filled with oil and the transformer is hermetically sealed.

After impregnation, the instrument transformers are also left for a few days before the partial discharge measurement and dielectric tests are carried out.

Since the impregnation in this process takes place at room temperature and therefore the viscosity of the oil is higher than when impregnated in an oven, the instrument transformer is pressurised for some time after the oil filling (step 3).

Disadvantages of this process are the possibility of reabsorption of moisture after drying during assembly of the instrument transformer and the higher viscosity of the oil at room temperature. Experience has shown that this process leads to a higher failure rate during testing (partial discharge measurement) and should therefore be avoided if possible.

The advantages of this process are the smaller space requirement and the faster turnover in the vacuum vessel, as only the active parts are brought into the vessel and for the time of impregnation the vessel can already be used again for the next active parts.

The process a. with drying and impregnation of the assembled instrument transformer in a vacuum vessel has proved to be the better and more reliable method. Table 3.2 shows a comparison of the two processes.

3.2.9   Temperature Rise of the Current Transformer

3.2.9.1   Requirements from the Instrument Transformer Standard for Maximum Temperatures and Temperature Rise in the Current Transformer

An increased temperature in the instrument transformer accelerates the ageing of the insulation. To ensure safe operation of the instrument transformer, the temperatures in the current transformer and its connections must not become too high. The instrument transformer standard IEC 61869-1 specifies maximum temperature rise which must not

Table 3.2 Comparison of the two processes for drying and impregnation

Drying and impregnation Process a.	Drying and impregnation process b.
Active part is built into the housing	–
Instrument transformer goes into vacuum vessel	Active part goes into vacuum vessel
Heating of the vessel (70 °C, 800 mbar)	Heating of the vessel (70 °C, 800 mbar)
Pressure reduction (30 mbar)	Pressure reduction (30 mbar)
Intermediate ventilation and further pressure reduction (up to 120 °C can be achieved)	Intermediate ventilation and further pressure reduction (up to 120 °C can be achieved)
Vacuum (120 °C, ≤0.04 mbar)	Vacuum (120 °C, ≤0.04 mbar)
Post-drying phase, 2–5 days depending on insulation thickness (120 °C, ≤0.01 mbar)	Post-drying phase, 2–5 days depending on insulation thickness (120 °C, ≤0.01 mbar)
Cooling down the boiler to 70 °C	Dried active part is installed in housing
–	Vacuum and oil line are connected at the filling station
–	Pressure reduction to 0.1 mbar for 48 h
Oil filling of the transformer with a quantity per hour depending on the size	Oil filling of the transformer with a quantity per hour depending on the size
Post-impregnation phase (12 h): Oil is topped up when the oil level drops	Pressure 2,6 bar to support the impregnation (up to 245 kV 24 h, >245 kV 72 h)
Opening the vessel and filling the bellows with oil	Filling the bellows with oil
Rest period before voltage test (4 to 8 days, depending on the size of the transformer)	Rest period before voltage test (4 to 8 days, depending on the size of the transformer)

be exceeded for the continuous thermal operating current (see Table 3.3). Unless otherwise specified, the continuous thermal operating current is 120% of the rated current.

The table is divided into 3 sections:

1. Values for oil-insulated instrument transformers, including toroidal core current transformers which are operated in oil (e.g., as bushing type current transformers in power transformers)
2. Values for solid-insulated and gas-insulated instrument transformers. Depending on the temperature class according to the IEC 60085 standard [44] of the insulating material, a distinction is made.

Table 3.3 Maximum permissible temperatures and temperature increases in transformers with different insulation materials. (from IEC 61869-1 [33])

Part of the instrument transformer	Maximum value for	
	Temperature in °C	**Temperature rise** at a maximum ambient temperature of 40 °C in K
1) **Oil-insulated instrument transformer** • Oil on top of the IT • Oil on top of the IT if hermetically sealed • Average in the winding • Average in the winding if hermetically sealed • Other metallic parts in contact with oil	90 95 100 105 Same as winding	50 55 60 65 Same as winding
2) **Solid-insulated or gas-insulated instrument transformers** • Average of the winding in contact with insulating materials of the following classes (°C)[b]: - 90 (Y) - 105 (A) - 120 (E) - 130 (B) - 155 (F) - 180 (H) • Other metallic parts in contact with insulating materials of the above classes	85 100 115 125 150 175 Same as winding	45 60 75 85 110 135 Same as winding
3) **Screwed or equivalent connections (including terminals)** bare copper or aluminium alloys • in air • in SF$_6$ • in oil silver-plated or nickel-plated • in air • in SF$_6$ • in oil tin-coated • in air • in SF$_6$ • in oil	90 115 100 115 115 100 105 105 100	50 75 60 75 75 60 65 65 60

b) according to IEC 60085

3. Values for screwed or similar connections in and on the instrument transformer. This also includes the terminals of the instrument transformer. With these connections, a distinction is made between bare copper or aluminium alloys and surface-treated connections.

If a temperature range exceeding +40 °C is specified, the maximum permissible temperature rise must be reduced accordingly.

The locations listed in the table are shown in Fig. 3.50.

The temperature of the windings is an average value and is determined by measuring the resistance. The maximum values for the various insulating materials have been reduced by 5 °C compared to the standard IEC 60085 [44] to consider that the actual maximum temperature is not measured by determining the average value.

3.2.9.2 Consideration of the Installation Altitude of the Instrument Transformer

The values given in Table 3.3 apply up to an installation altitude of 1000 m above sea level. If a transformer is operated at higher altitudes, the maximum permitted temperature rise is reduced by 0.4% for oil-insulated transformers and 0.5% for solid and gas-insulated transformers for each 100 m that the altitude exceeds 1000 m. The maximum permitted temperature rise at an installation altitude h, ΔT_h, is calculated as follows: $\Delta T_h = K_0 \cdot \Delta T_{h0}$ where K_0 is the correction factor and ΔT_{h0} is the permitted temperature rise at an installation altitude of up to 1000 m (according to Table 3.3).

The factor K_0 is shown in Fig. 3.51.

For example, if a hermetically sealed oil-insulated instrument transformer is operated at an altitude of 2000 m above sea level, the permitted average temperature increase of the windings is

Fig. 3.50 Temperature measurements on an oil-insulated current transformer

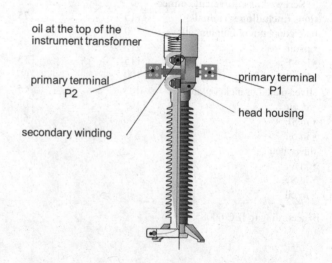

oil at the top of the instrument transformer

primary terminal P2

primary terminal P1

head housing

secondary winding

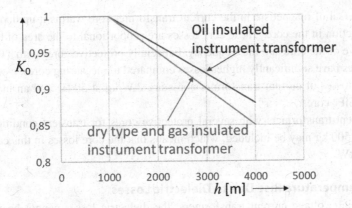

Fig. 3.51 Altitude correction factor for maximum temperature rise

$$\Delta T_h = K_0 \cdot \Delta T_{h0} = 0,96 \cdot 65\,K = 62.4\,K$$

(with $\Delta T_{h0} = 65$ K according to Table 3.3 and K0 = 9.96 according to Fig. 3.51).

3.2.9.3 Temperature Rise by the Operating Current

During operation of the current transformer, the following losses contribute to the temperature rise in the current transformer:

a) Primary current in the primary conductor

In Sect. 3.2.7.2 the losses due to the operating current in the primary conductor were calculated to $P_P = I_P^2 \cdot R_P$. These depend on the cross-section of the primary conductor and its length. As shown in Sect. 3.2.7.2, a hairpin current transformer produces significantly higher losses because of its longer primary conductor. In the example shown these were 1235 W in a hairpin currant transformer, compared to 103 W in a head type current transformer.

With several primary turns, the losses in the primary conductor are increased because of the usually smaller conductor cross-sections. Especially with many primary windings as shown in Fig. 3.37 the allowed primary current is limited because of the high losses.

b) Secondary current in the secondary winding

The current in the secondary winding heats it and thus contributes to the temperature rise of the current transformer. The losses are calculated as $P_S = I_S^2 \cdot R_S$.

With typical winding resistances of 5 Ω and a secondary current of 1 A, the losses are 5 W, but if a secondary current of 5 A is specified, the losses increase significantly to 125 W per transformer core.

c) Core losses in the current transformer core

The reversal of magnetism in the current transformer core warms it up depending on the induction in the core. These core losses are proportional to the area of the hysteresis curve of the core and the frequency. Especially protective cores with TPY or TPZ properties have significantly higher losses compared to measuring cores.

Typical losses of current transformer cores are 2 W/kg at 50 Hz and an induction of 1 T for SiFe cores.

For current transformers with several protective cores for transient conditions, cores of up to 500 kg may be included, which means that the core losses in this case can be up to 1 kW.

3.2.9.4 Temperature Rise Due to Dielectric Losses

For extra-high voltage current transformers, the dielectric losses cannot be neglected. The standard IEC 61869-2 therefore requires that for current transformers with U_m greater than or equal to 550 kV, the temperature rise test must be carried out simultaneously with the thermal continuous current and the maximum operating voltage $U_m/\sqrt{3}$.

The dielectric losses of a transformer with $U_m = 550$ kV and a loss factor $tan\delta = 0.003$ with a typical capacity of the current transformer of 1 nF are at 50 Hz:

$$P_\delta = \left(U_m \big/ \sqrt{3}\right)^2 \cdot \omega \cdot C \cdot \tan \delta = 95 \, W.$$

3.2.9.5 Temperature Rise Due to the Short-Circuit Current

If a short-circuit current flows through the transformer, both the primary conductor and the secondary winding will heat up additionally. The standard IEC 61869-2 does not specify any limits for this, but after the short-circuit test, no carbonisation may occur when the insulation is examined. This examination is not necessary if the current densities in the primary and secondary winding are not greater than 180 A/mm^2 for copper and 120 A/mm^2 for aluminium.

With a short-circuit current of 1 s duration, adiabatic heating can be assumed. In general, the following applies to adiabatic heating:

$$\Delta T = \frac{P \cdot t}{c \cdot m}$$

where P is the power in W=VA, t the time in sec, c the specific heat of the conductor material in Ws/gK and m the mass in g.

The power P, which is converted in the conductor, is:

$P = I^2 \cdot R = j^2/_{A^2} \cdot \rho^l/_A$ where j is the current density I/A and R is the resistance of the conductor, l is the length of the conductor in m, A is the cross-section of the conductor in mm^2 and ρ is the resistivity of the conductor material in Ωmm/m.

Example 1: For an aluminium primary conductor with a minimum cross-section for 40 kA (120 A/mm^2)

Short-circuit current $\quad I = 40\,000$ A

Cross section $\qquad\quad A = 333$ mm^2

Conductor length $\quad l = 0.5$ m

specific heat $\qquad\quad c = 0.946$ Ws/gK

specific resistance $\quad \rho = 27.78\ 10^{-3}$ Ωmm/m

specific weight $\qquad \gamma = 2.7$ g/cm^3

Thus, the temperature rise for a short circuit duration of 1 s:

$$\Delta T = \frac{I^2 \rho {}^l\!/_A \cdot t}{c \cdot m} = \frac{40,000^2 \cdot 27.78 \cdot 10^{-3} \cdot 0.5/333 \cdot 1}{0.946 \cdot 2.7 \cdot 50 \cdot \pi \cdot 3.33} = 59.96\,K$$

Example 2: For a copper primary conductor with a minimum cross-section for 40 kA (180 A/mm2)

Short circuit current $\quad I = 40,000$ A

Cross-section $\qquad\quad A = 222$ mm^2

Conductor length $\quad l = 0.5$ m

specific heat $\qquad\quad c = 0.386$ Ws/gK

specific resistance $\quad \rho = 17.86\ 10^{-3}$ Ωmm/m

specific weight $\qquad \gamma = 8.96$ g/cm^3

Thus, the temperature rise for a short circuit duration of 1 s:

$$\Delta T = \frac{I^2 \rho {}^l\!/_A \cdot t}{c \cdot m} = \frac{40,000^2 \cdot 17.86 \cdot 10^{-3} \cdot 0.5/222 \cdot 1}{0.386 \cdot 8.967 \cdot 50 \cdot \pi \cdot 2.22} = 53.32\,K$$

3.2.9.6 Examples from Temperature Rise Tests

The results of a temperature rise test are shown below.

The current transformer of the example of type IOSK 245 from Trench had the following data:

highest voltage U_m	245 kV
rated primary current I_{pr}	2000 A
continuous thermal current I_{th}	3000 A
rated secondary current I_{sr}	1 A
rated frequency f_r	50 Hz
Core 1	Class 0.2S, 15 VA
Cores 2–5	Class 5P, 30 VA

A primary current of 3000 A flowed through the primary conductor until the temperatures in the transformer stabilised. Figure 3.52 shows the schematic diagram of the measurement with the supply of the primary current and the position of the temperature measuring positions on the current transformer.

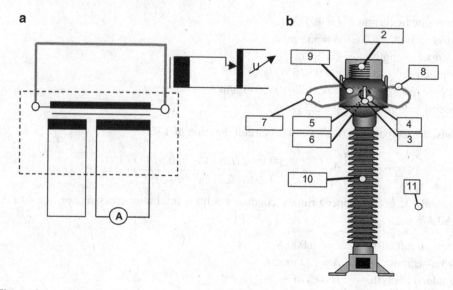

Fig. 3.52 a Schematic diagram of the test **b** Position of the temperature measurements

Table 3.4 shows the measured temperatures at the beginning of the test and after the test. According to IEC 61869-2, the test ends if the temperature rise is less than 1 degree in one hour.

The secondary windings were loaded with the rated burden. The temperature rise of the secondary windings was determined by the change of the winding resistance. Table 3.5 shows the temperature rise of the secondary windings.

3.2.10 Mechanical Stress

3.2.10.1 Forces Acting on the Current Transformer

The following forces can act on the current transformer during operation:

- Forces due to line connection at the primary terminals F_S
- Wind forces F_W
- Forces due to short-circuit currents F_K
- Seismic forces F_E

The terminal forces and wind forces are static forces and can load the instrument transformer for a long time, whereas short-circuit forces and seismic forces are dynamic forces and only occur for a short time.

Table 3.4 Measured temperatures before and after the temperature rise test

Measuring point	Position	Temp. before the test	Temp. after the test		Temperature rise
		Time t_0 (°C)	Time t_1 (°C)	Time $t_1 + 1$ h (°C)	(K)
2	Bellows, top	25.5	38.5	38.9	13.4
3	Primary termi-nal P1	25.5	63.9	64.4	38.9
4	primary conductor	25.5	57.7	58.3	32.8
5	Primary termi-nal P2	25.5	60.4	61.2	35.7
6	Primary conductor	25.5	55.1	55.8	30.3
7	Connection cable 1 m from P1	25.5	69.0	68.8	43.3
8	Connection cable 1 m from P2	25.5	65.0	65.1	39.6
9	Head housing	25.5	37.7	38.2	12.7
10	Insulator	25.5	33.9	34.3	8.8
11	Ambient temperature	25.5	25.6	25.5	0.0

Table 3.5 Temperature rise of the secondary windings

Core No	R_0 in Ω before the test	R_T in Ω at the end of the test	Average temperature rise (K)
1	4.632	5.572	52.8
2	9.281	11.224	54.4
3	9.512	11.544	55.5
4	9.316	11.299	55.3
5	8.434	10.169	53.5

The critical area of the mechanical load on the instrument transformer is the cementing area of the insulator in the lower flange. The maximum breaking moment at this point is the sum of the moments of the individual forces. Figure 3.53 shows the points of application of the forces on a current transformer.

Fig. 3.53 Forces on the current transformer Line connection and short-circuit force at primary terminal $F_A = F_S + F_K$; Wind force at the surface centre F_W; Seismic force at the centre of gravity F_E

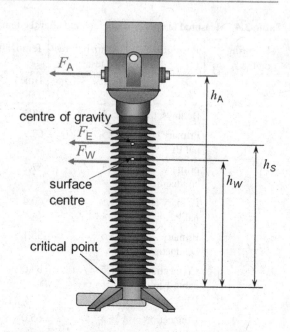

centre of gravity F_E

F_W

surface centre

critical point

In the following the forces and their occurring values are described:

a) Forces due to line connection at the primary terminals
   The tensile force of the connected conductor permanently loads the terminals of the instrument transformer. It is determined by the operator for the respective installation in the switchgear. It is typically 500 to 2000 N. The direction of force is horizontal in the direction of the conductor, but due to the weight of the conductor part of the force will also act vertically downwards.
   To calculate the moment M_S on the insulator at the cementing area, the tensile force F_S is multiplied by the height of the primary connection above the cementing area.
   $M_S = h_A \cdot F_S$,
   where h_A is the height of the primary conductor above the cementing area.

b) wind force:
   Wind acts on the entire contact surface of the instrument transformer. The force F_W on the instrument transformer is calculated by:

$$F_W = 0.5 \cdot \rho_L \cdot v_W^2 \cdot A$$

Here ρ_L is the air density (1.3 kg/m³), v_W is the wind speed in m/s and A is the contact surface of the instrument transformer in m².
For an example of a 245 kV current transformer, a wind speed of 30 m/s (corresponding to 108 km/h) and a contact surface of 0.72 m² results in a wind force of 421 N.
To calculate the moment due to the wind force at the cementing area, the total wind force is applied at the centre of the surface. The moment due to the wind is thus calculated as

$$M_W = h_W \cdot F_W,$$

where h_W is the height of the surface centre above the cementing area.

c) Forces due to short-circuit current

In addition to the permanent tensile force, the short-circuit force generated by the high currents in the supply connector acts at the primary terminal in the event of a short circuit.

The short-circuit force acts for a short time as a dynamic force and has its peak as a force impulse in the peak of the dynamic short-circuit current.

The force between two conductors during the short circuit can be calculated:

$$F_K = \mu_0 \cdot \frac{I_{dyn}^2}{2\pi} \cdot \frac{l}{a}$$

Here μ_0 is the permeability of the air between the conductors ($\mu_0 = 4\pi 10^{-7}$ Vs/Am), I_{dyn} is the dynamic peak of the short-circuit current, l is the length of the conductor section under consideration and a is the distance between the conductors. For the case of forces at the current transformer this is the distance between the phases.

For an example of a 245 kV transformer with short-circuit current 40 kA ($I_{dyn} = 2.5 \times 40 = 100$ kA) and a phase distance of 5 m with an effective conductor length of 2 m, this results in a peak force of 800 N.

d) Seismic Forces

Another short-term dynamic force is exerted on the instrument transformer by earthquakes. The acceleration at ground level a during an earthquake is increased by the increased support structure of the instrument transformer. Usually, amplification factors k_a of 2 to 4 are used for this purpose.

The moment M_E at the cementing area due to the seismic load is calculated:

$$M_E = h_S \cdot a \cdot k_a \cdot m,$$

where h_S is the height of the centre of gravity above the cementing area and m is the mass of the instrument transformer.

However, this static calculation of the seismic forces only shows the actual load of an instrument transformer during an earthquake in a first approximation.

More precise simulations of the dynamic conditions can be calculated with the help of computer programs and thus the load on the instrument transformers can be predicted. The calculations can be confirmed by seismic tests in which the instrument transformer is exposed to an artificial earthquake on a shaking table. The necessary amplification factors can also be determined. For these tests, the instrument transformer must be mounted on its support structure, as this also has an influence on the amplification factors. Figure 3.54 shows a current transformer during an earthquake test.

Since the seismic forces can occur in any direction, a symmetrical support structure of the instrument transformer is important. The best results are achieved if the support is made by a tube.

Fig. 3.54 Experimental set-up
of a seismic test with a 145 kV
current transformer type IOSK
145 of the company Trench
in the laboratory of ISMET in
Italy

3.2.10.2 Strength of the Insulators

As already described in Sect. 1.7.3 and 1.7.4 the strength of the insulators depends on
the material and the cementation. The insulator manufacturers specify guaranteed breaking forces for the insulators used. The maximum moment at the cementing area is therefore $M_P = F_P h_P$, with the guaranteed cantilever force F_P and the height of the insulator h_P
above the cementing area.

The moments described above affect at the place of cementation.

To ensure safe operation of the instrument transformers, the safety factor.

$SF = M_P/(M_S + M_W + M_E)$ must not be less than 2.

3.2.10.3 Requirements and Tests According to the Instrument
Transformer Standard

The instrument transformer standard IEC 61869-1 requires mechanical strength for all
instrument transformers. It requires a static force to be applied to the primary connections which depends on the voltage level and the load class. The force must be applied to
the terminals in all three directions for 60 s each. This must not lead to any destruction or
leakage.

3.3   Low-Power Current Transformer (Iron Core with Integrated Measuring Shunt)

3.3.1   Introduction

With the development of new protection systems, meters and monitoring devices based on microprocessor technology, the power requirement of the current transformer output has decreased by a factor of 100 to 1000, so that the current transformers used for this purpose can be designed for this power requirement. The necessary guidelines are laid down in the IEC standard IEC 61869-10 [37].

3.3.2   The Principle of the Low-Power Current Transformer with Integrated Measuring Shunt

The European patent EP0990160B [45] describes a current transformer with voltage output which is practically immune to electrical and magnetic interference.

The two connection circuits of standard current transformers in Fig. 3.55 and 3.56 show the areas of the connection systems where interference can be induced by external fields.

Figure 3.55 shows the circuit of a standard current transformer with a burden R_b. The output voltage across the burden is transmitted to the connected devices via transmission cable 2.

Figure 3.56 shows the standard connection system of the current transformer. The burden R_b is the sum of the parallel connected input resistances R_e of the protection and measuring devices.

In the circuit with integrated shunt as described in the above-mentioned patent, the measurement of the current is achieved by forming a part of the winding as a shunt with

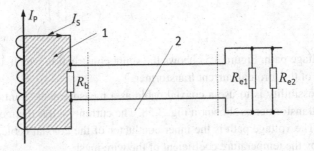

Fig. 3.55 Connection of a standard current transformer with burden R_b and transmission cable 2. The shaded area 1 represents the area where interference voltages can be induced. (I_P—Primary current; I_S—Secondary current; R_b—burden in the terminal box of the instrument transformer; R_{e1}, R_{e2}—input resistors of connected devices)

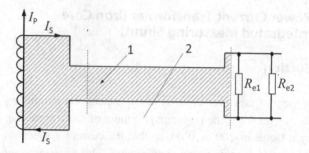

Fig. 3.56 Standard connection system wiring. The shaded area 1 represents the area where voltages can be induced. The connecting cable 2 is part of the burden of conventional current transformers. (I_p—Primary current; I_s—Secondary current; R_{e1}, R_{e2}—Input resistances of connected devices)

Fig. 3.57 Wire shunt before twisting the resistance wires

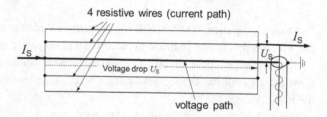

Fig. 3.58 Coaxial shunt which is also part of the winding of a toroidal current transformer. At the beginning of the measurement shunt, the winding mesh and conductor are connected together

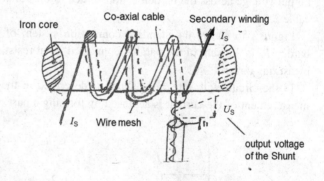

current and voltage path. Figure 3.57 shows the shunt construction which, twisted, is part of the winding of the toroidal current transformer.

A second possibility is to use a coaxial cable as a measurement shunt in a short-circuited current transformer as shown in Fig. 3.58. The current path is the mesh wire of the coaxial cable. The voltage path is the inner conductor of the coaxial cable. The stability is determined by the temperature coefficient of the wire mesh.

3.3.3   Determination of the shunt resistance R_{Sh}

The systems commonly used today in power engineering process signals from $\leq \pm 10$ V.
   The shunt resistance is determined by:

$$R_{Sh} \leq \frac{U_S}{2 \cdot \sqrt{2} \cdot I_{th} \cdot \ddot{u}}$$

where

R_{Sh} resistance of the shunt
U_S output voltage of the shunt
I_{th} thermal short-circuit current of the current transformer
$\ddot{u}$ transmission ratio of the current transformer

Example of a 10 V output signal with a current of 10 kA and a transformer ratio of the
current transformer of 1/3000

$$R_{Sh} \leq \frac{10\,\text{V}}{2 \cdot \sqrt{2} \cdot 10{,}000\,\text{A} \cdot \frac{1}{3000}} = 1.1\,\Omega$$

3.3.4   The Theory of the Low-Power Current Transformer with Integrated Measurement Shunt

3.3.4.1 Comparison of Technologies for Current Transformers

Table 3.6 compares the three basic technologies for current transformers and their characteristics [46].
   Technology T1 is the conventional current transformer and is described and analysed in Sect. 3.2. This Sect. 3.3 deals with technology T3, the low-power current transformer with integrated shunt, and Sect. 3.4 deals with technology T2, the air core coil
(Rogowski coil).

3.3.4.2 Error of the Low-Power Current Transformer

a. The amplitude error
   Figure 3.59 shows the equivalent circuit diagram of the closed toroidal current transformer with integrated shunt and voltage output.
   The equivalent circuit diagram is related to the primary side. The current transformer
can also be operated potential-free.
   P1 and P2 are the primary terminals and S1 and S2 are the secondary terminals of the
current transformer with voltage output.
   The current/voltage transformer is the shunt ($R_{Sh} \cdot \ddot{u}^2$), which converts the current signal ($I_z/\ddot{u}$) into the voltage signal ($U_s \cdot \ddot{u}$).

Table 3.6 Current transformer technologies

No	Technology	T1 Iron core, output 1A or 5 A state of the art	T2 air-core coil with voltage output (Rogowski coil)	T3 Universal toroidal current transformer with voltage output
	Standard:	IEC 61869-2	IEC 61869-10	IEC 61869-10
	Properties:			
1	Open current transformer output	(−) dangerous, the transformer must not be operated with open secondary windings	(++) Secondary side can be open	(++) Secondary side can be open
2	Universal current transformer for measurement, protection and good transient characteristics	(−) requires separate cores for measurement and protection	(+) sufficient for certain areas of measurement and well suited for protection purposes	(++) applicable for current ranges from 50 A to 25 kA/63 kA Measurement and protection
3	Burden area	(+) within the specified burden range	(−) Calibration required for each burden	(**) for all burdens >20 kΩ without calibration
4	Phase shift	(**) $\Delta\varphi < 5'$ to $60'$	(++) physically caused phase shift of $-90°$	(++) $\Delta\varphi < 5'$ to $60'$
5	Losses for a connection of 300 m length and a conductor cross-section of 6mm2	(−) 1 A Output: 5 VA (−) 5 A Output: 25 VA	(++) negligible, because voltage output	(++) negligible, because voltage output
6	Transient behaviour in protective applications	(−) requirements can be solved by costly solutions (by using linearised cores)	(++) linearised transmission behaviour of the air core coil (−) high current changes cause high voltage peaks, $U = Ldi/dt$	(++) Due to small burden R_b and high number of turns, core saturation only occurs at high short-circuit current

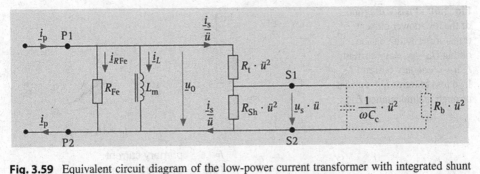

Fig. 3.59 Equivalent circuit diagram of the low-power current transformer with integrated shunt [46] (I_P—primary current; I_{Rfe}—current through the iron loss resistor; R_{Fe}—equivalent iron loss resistance; L_m—Main inductance; $I_S/\ddot{u}$—Secondary current related to the primary side ($\ddot{u} = N_P/N_S$); U_0—Equivalent voltage; $R_t \cdot \ddot{u}$—Winding resistance related to the primary side)

From Fig. 3.59 the amplitude error ε is derived if I_L is neglected:

$$\varepsilon = \frac{\left|\frac{I_S}{\ddot{u}}\right| - |I_P|}{|I_P|} = \frac{\frac{U_0}{(R_t + R_{Sh})\ddot{u}^2}}{\frac{U_0}{(R_t + R_{Sh})\ddot{u}^2} + \frac{U_0}{R_{Fe}}} - 1$$

$$\varepsilon = \frac{1}{1 + \frac{(R_t + R_{Sh})\ddot{u}^2}{R_{Fe}}} - 1$$

By a series development broken off after the first link results:

$$\varepsilon = 1 - \frac{(R_t + R_{Sh})\ddot{u}^2}{R_{Fe}} - 1 \approx -\frac{(R_t + R_{Sh})\ddot{u}^2}{R_{Fe}}$$

With this the amplitude error is:

$$\varepsilon \approx -\frac{(R_t + R_{Sh})\ddot{u}^2}{R_{Fe}}$$

The amplitude error is determined by the resistances of the secondary winding R_t and the shunt R_{Sh}, which is integrated in the winding, at a given R_{Fe}. The connected burden R_b for $R_b > 20\,k\Omega$ has no measurable influence on the accuracy.

b. *The phase displacement* $\Delta\varphi$

From Fig. 3.59 the phasor diagram in Fig. 3.60 is derived

This results in a phase shift $\Delta\varphi$ between primary current I_P and secondary current I_S or secondary voltage U_S:

$$tan\Delta\varphi = \frac{|I_L|}{\left|I_S/\ddot{u}\right| + |I_{Rfe}|}$$

Fig. 3.60 Phasor diagram of the low-power current transformer with voltage output (I_P—primary current; $I_S/\ddot{u}$—secondary current; I_{Rfe}—Iron loss current; I_L—current through the inductance L_m)

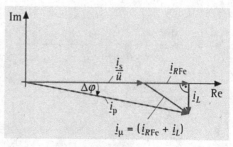

I_P	primary current
$I_S/\ddot{u}$	secondary current
I_{Rfe}	Iron loss current
I_L	current through the inductance L_m

With the equivalent voltage U_0 follows:

$$tan\Delta\varphi = \frac{U_0/\omega L_m}{\frac{U_0}{(R_t + R'_{Sh})\ddot{u}^2} + \frac{U_0}{R_{Fe}}} = \frac{(R_t + R'_{Sh})\ddot{u}^2}{\omega L_m \left(1 + \frac{R_t + R'_{Sh}}{R_{Fe}}\ddot{u}^2\right)}$$

R'_{Sh} is the parallel connection of the shunt resistor R_{Sh} and the connected burden R_b:

$$R'_{Sh} = \frac{R_{Sh} \cdot R_b}{R_{Sh} + R_b} = \frac{R_{Sh}}{1 + R_{Sh}/R_b}$$

With $R_{Sh} \approx 1\,\Omega$ and $R_b \geq 20\,\text{k}\Omega$, $R'_{Sh} \approx R_{Sh}$ and with $\frac{R_t + R'_{Sh}}{R_{Fe}}\ddot{u}^2 \ll 1$ follows:

$$tan\Delta\varphi \approx \frac{R_t + R_{Sh}}{\omega L_m}\ddot{u}^2 \text{ or } \Delta\varphi \approx \frac{R_t + R_{Sh}}{\omega L_m}\ddot{u}^2$$

The positive phase angle $\Delta\varphi$ between secondary current I_S or secondary voltage U_S and primary current I_P increases with the sum of secondary winding resistance R_t and measuring shunt R_{Sh} and is reduced by a larger inductance L_m.

3.3.4.3 Influence of Temperature on the Current/Voltage Converter R_{Sh}

The resistance of the measurement shunt changes with the temperature:

$$R_{Sh} = R_{Sh20°C}(1 + \alpha\Delta\vartheta)$$

For the amplitude error of the low-power current transformer $\varepsilon \approx -\frac{(R_t + R_{Sh})\ddot{u}^2}{R_{Fe}}$ a rising temperature $\Delta\vartheta$ results in a larger negative current error if the temperature coefficient α is positive [47].

The accuracy of the low-power current transformer is determined by the temperature coefficient of the shunt R_{Sh}, which acts as a converter that converts the current into a voltage. The temperature coefficient α is clearly defined as a function of temperature by the tangent to the resistance curve for a certain temperature T of the conductor material and is:

$$\alpha_x = \frac{\left(\frac{\partial R}{\partial T}\right)_{T=T_x}}{R_{T=T_x}} \text{ in } 1/K$$

Depending on the temperature range, the temperature coefficient can be positive or negative. The change can be continuous or discontinuous, and at a temperature $T_x = T_0$, $\alpha = f(T) = f(T_0) = 0$.

The value T_0 changes from batch to batch during the production of the resistance alloy and can be considered when designing the shunt. Only with practical knowledge of resistor technology the shunts can be built and integrated into the current transformer with $\alpha \le \pm 5 \cdot 10^{-6}$ 1/K.

By selecting the characteristic of $R_{Sh} = R_{Sh20°C}(1 + \alpha \Delta \vartheta)$ a very small temperature influence of the error can be achieved.

3.3.4.4 The Equivalent Voltage Source U_E of the Low-Power Current Transformer with Integrated Shunt

The voltage at the output of the low-power current transformer with integrated shunt is calculated from the diagram in Fig. 3.61.

An assessment of the influence of the cable resistance R_l and the burden resistance R_b on the accuracy of a standardised voltage interface between the current transformer and the secondary devices is provided by the equivalent voltage source of the low-power current transformer in Fig. 3.61.

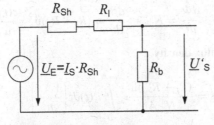

I_s	Secondary current
R_{Sh}	Shunt integrated in the winding
U_E	equivalent voltage source
U'_s	Voltage at burden R_b
R_l	resistance of the connecting cable between current transformer and secondary devices
R_b	sum of the parallel connected input resistances of secondary devices

Fig. 3.61 Equivalent voltage source of the low-power current transformer

The error for the measurement is derived from the Figure to:

$\varepsilon = \frac{U'_S - U_E}{U_E}$ with $U'_S = \frac{R_b}{R_{SH} + R_l + R_b} \cdot I_S \cdot R_{Sh}$ and $U_E = I_S \cdot R_{Sh}$ it results in:

$$\varepsilon = \frac{\frac{R_b}{R_{Sh} + R_l + R_b} \cdot I_S \cdot R_{Sh} - I_S \cdot R_{Sh}}{I_S \cdot R_{Sh}} = -\frac{R_{Sh} + R_l}{R_{Sh} + R_l + R_b} = -\frac{R_{Sh} + R_l}{R_b \left(1 + \frac{R_{Sh} + R_l}{R_b}\right)}$$

With $\frac{R_{Sh} + R_l}{R_b} \ll 1$ the error results in:

$$\varepsilon = -\frac{R_{Sh} + R_l}{R_b}$$

The value for R_b is standardised in the IEC 61869-10 standard at 2 MΩ. In the earlier standard IEC 60044-8, smaller values, 20 kΩ and 2 kΩ were also standardised.

For the consideration of the error, 2 cases must be distinguished, depending on the cable length between transformer and secondary equipment:

a) For medium voltage and GIS systems (cable length approx. 6 m) $R_{Sh} \approx 1\Omega \gg R_l$. For $R_b \geq 2$ kΩ follows: $\varepsilon \approx -\frac{1}{2000} = -0.5 \cdot 10^{-3} = -0.05\%$

b) In open air systems (cable length 300 m) with $R_l = 2\Omega$ is for $Rb \geq 2$ kΩ $\varepsilon \approx -\frac{1+2}{2000} = -1.5 \cdot 10^{-3} = -0.15\%$

3.3.4.5 The Transient Behaviour of the Low-Power Current Transformer

The relationship between the time course of the short-circuit current, the current transformer parameter and the saturation flux density B_{max} in the iron core of the ring current transformer [48], which still allows a processing of the primary current $i_p(t)$ that can be processed by the secondary technology, follows from the equivalent circuit in Fig. 3.59:

$$U_0(t) = N \cdot \frac{d\Phi}{dt} = N \cdot A \cdot \frac{dB}{dt} = (R_t + R_S)\ddot{u}^2 \cdot \frac{i(t)}{\ddot{u}}$$

The following applies to the flux density:

$$B(t) = \frac{(R_t + R_{Sh})\ddot{u}}{N \cdot A} \cdot \int_{t=0}^{t=tx} i_s(t)dt \left[\frac{V \cdot s}{m^2}\right]$$

where $B(t) \leq$ is B_{max}.

The permissible value of B_{max} is determined by the magnetisation curve of the iron core. B_{max} is approximate (for silicon iron cores):

$$B_{max} < 1.8 \ldots 2\,Vs/m^2$$

In the equation $B(t)$ and $i_S(t)$ are given by the primary current. The time t_x determines the time at which the transformer reaches the saturation flux density.

The expression $\frac{(R_t + R_{Sh})\ddot{u}}{N \cdot A}$ (sum of the winding resistance of the secondary winding R_t and the resistance of the shunt R_{Sh}, divided by the product of the number of turns N

of the transformer multiplied by the area A of the iron core), together with the current integral, must meet the requirements for transient behaviour of the current transformer according to the standard IEC 61869-2. The smaller the expression, the better the transient behaviour.

3.3.5 Practical Versions of the Low-Power Current Transformer

The two Figs. 3.62 and 3.63 show practical examples of the universal low-power current transformer with voltage output. Figure 3.62 shows a low-power current transformer for installation in GIS systems. It is suitable for primary currents up to 5000 A and has a weight of 5.5 kg. The diameter is adapted to the GIS system.

Figure 3.63 shows a low-power current transformer with voltage output for installation in a cable outlet to measure the currents in the cable.

Fig. 3.62 Low-power current transformer with voltage output for GIS application (type LPCT25 C [49])

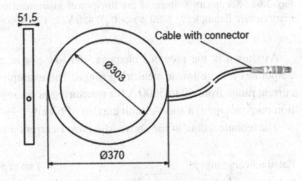

Fig. 3.63 Low-power current transformer with voltage output for installation on the cable outlet (type LPCT 25 A [49])

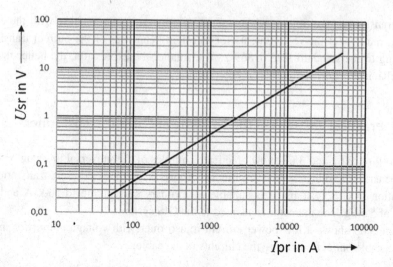

Fig. 3.64 Secondary voltage of the low-power instrument transformer as a function of the primary current Example: I_p 1000 A => U_S 0.450 V; I_p 10 kA => U_S 4.50 V

As shown in the previous chapters, the low-power current transformer with voltage output has a large linear measuring range. The examples shown above can be used in a current range from 50 to 5000 A for measuring purposes and at the same time for protection purposes up to a short-circuit current of 63 kA.

The technical data of the above low-power current transformer [49]:

Rated primary range:	50 A up to 5000 A
Secondary voltage:	22.5 mV up to 2.25 V
Nominal frequency:	16.7 Hz, 50 Hz and 60 Hz
Short circuit current:	63 kA/3 s
Accuracy measurement:	class 0.2, 0.5 or 1
Accuracy protection:	5P up to 63 kA
Burden:	≥ 20 kΩ

Figure 3.64 shows the linear relationship between secondary voltage U_S and primary current I_p over a wide current range from 50 A to 63 kA.

3.4   Air Core Coil/Rogowski Coil

3.4.1   The Theory of the Air Core Coil

For the calculation of the induced voltage the following conductor coil configuration is used as a basis:

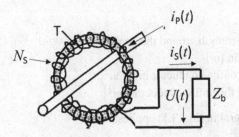

Fig. 3.65 The principle of the air core coil ($i_p(t)$—Primary current; N_s—Winding on non-magnetic core; $U(t)$—Voltage at burden Z_b; T—Non-magnetic core (air-core); Z_b—Burden of the air core coil; $i_s(t)$—Secondary current)

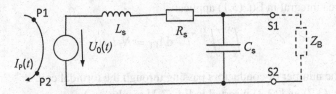

$I_P(t)$	Primary current
$U_0(t)$	induced voltage
L_s	Stray inductance of the air core coil
R_s	Resistance of the secondary winding
C_s	Capacity of the winding (is neglected)

Fig. 3.66 Air core coil equivalent circuit diagram

From the principle shown in Fig. 3.65 the equivalent circuit diagram in Fig. 3.66 follows

3.4.1.1 Calculation of the Induced Voltage $U_0(t)$

For the induced voltage $U_0(t)$ applies:

$$U_0(t) = M \cdot \frac{\partial i_p(t)}{\partial t}$$

M is the mutual inductance of the air core coil.

For the voltage induced by the primary current $i_p(t)$, Maxwell's equations apply for a toroidal coil with any cross-section.

The 1st Maxwell's law in integral form says:

$$\oint \vec{H} d\vec{s} = \int_{x_0}^{x_1} \int_{y_0}^{y_1} \vec{j} \cdot d\vec{A_{F1}} \tag{3.1}$$

with:

$\vec{H}$ magnetic field strength around the conductor in $[A/m]$
$\vec{ds}$ length element in [m]
$\vec{j}$ current density of the conductor in $[A/m^2]$
$\vec{A}_{F1}$ cross section of the entire conductor $[m^2]$

For the rotational integral in Eq. (3.1) applies:

$$\oint \vec{H}\vec{ds} = \left|\vec{H}\right| \cdot 2\pi r \tag{3.2}$$

With r as the radius of the circle around the conductor in m on which the field strength H is calculated.

For the area integral in Eq. (3.1) applies:

$$\int_{x_0}^{x_1} \int_{y_0}^{y_1} \vec{j} \cdot d\vec{A}_{F1} = N_P I_P(t) \tag{3.3}$$

With N_p as the number of conductors passing through the toroidal coil.

From Eqs. (3.2) and (3.3) inserted in Eq. (3.1) results

$$\left|\vec{H}(t)\right| = \frac{N_P}{2\pi r} I_P(t) \tag{3.4}$$

The 2nd Maxwell's law in integral form is

$$\oint \vec{E}\vec{ds} = \mu_0 \cdot \mu_r \cdot \int_{x_0}^{x_1} \int_{y_0}^{y_1} \frac{d\vec{H}(t)}{dt} \cdot d\vec{A}_{F2} \tag{3.5}$$

where:

$\vec{E}$ electric field strength in V/m
μ_0 Permeability of the vacuum in Vs/Am
μ_r relative Permeability of core material
A_{F2} cross section of the core

If the toroidal coil has N_S windings, this applies:

$$N_S \cdot \oint \vec{E}\vec{ds} = U_0(t) = N_S \cdot \mu_0 \cdot \mu_r \cdot \int_{x_0}^{x_1} \int_{y_0}^{y_1} \frac{d\vec{H}(t)}{dt} \cdot d\vec{A}_{F2} \tag{3.6}$$

If Eq. (3.4) is inserted into Eq. (3.6), the result is:

$$U_0(t) = \frac{N_S \cdot N_P \cdot \mu_0 \cdot \mu_r}{2\pi} \cdot \int_{x_0}^{x_1} \int_{y_0}^{y_1} \frac{\vec{1}}{r} \cdot \frac{\partial I_P}{\partial t} \cdot d\vec{A}_{F2} \tag{3.7}$$

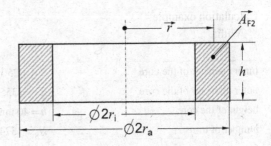

Fig. 3.67 Dimensions of an air core coil with rectangular core

If a rectangular core is defined for the coil according to Fig. 3.67, the area integral is calculated to:

$$U_0(t) = \frac{N_S \cdot N_P \cdot \mu_0 \cdot \mu_r}{2\pi} \cdot \frac{\partial I_P}{\partial t} \cdot \int_{r_i}^{r} \int_{y=0}^{y=h} \frac{\vec{1}}{r} \cdot d\vec{A_{F2}}$$

$$U_0(t) = \frac{N_S \cdot N_P \cdot \mu_0 \cdot \mu_r}{2\pi} \cdot \frac{\partial I_P}{\partial t} \cdot [lnr]_{r_i}^{r_a} \cdot h$$

$$U_0(t) = \frac{N_S \cdot N_P \cdot \mu_0 \cdot \mu_r}{2\pi} \cdot \ln\frac{r_a}{r_i} \cdot h \cdot \frac{\partial I_P}{\partial t} \tag{3.8}$$

The expression $\frac{N_S \cdot N_P \cdot \mu_0 \cdot \mu_r \cdot h}{2\pi} \cdot \ln\frac{r_a}{r_i}$ is called mutual inductance M.

For an air-core coil with a rectangular core cross-section (see Fig. 3.67) and one primary conductor, the following applies to the mutual inductance:

$$M = \frac{N_S \cdot \mu_0 \cdot h}{2\pi} \cdot \ln\frac{r_a}{r_i} \tag{3.9}$$

where $Np = 1$, $\mu_r = 1$. With $\mu_0 = 4\pi 10^{-7}$ Vs/Am follows from Eq. (3.9):

$$M = 2 \cdot N_S \cdot h \cdot \ln\frac{r_a}{r_i} \cdot 10^{-7} \frac{\text{Vs}}{\text{A}} \tag{3.10}$$

The induced voltage is thus:

$$U_0(t) = 2 \cdot N_S \cdot h \cdot \ln\frac{r_a}{r_i} \cdot 10^{-7} \frac{\text{Vs}}{\text{A}} \cdot \frac{\partial I_P}{\partial t} \tag{3.11}$$

With the primary current $i_P(t) = \hat{I} \cdot \sin \omega t$ follows from Eq. (3.11):

$$U_0(t) = 2 \cdot N_S \cdot h \cdot \ln\frac{r_a}{r_i} \cdot 10^{-7} \frac{\text{Vs}}{\text{A}} \cdot \omega \cdot \hat{I} \cdot \cos \omega t \tag{3.12}$$

And with the abbreviation Eq. 3.10:

$$U_0(t) = M \cdot \omega \cdot \hat{I} \cdot \cos \omega t \tag{3.13}$$

Calculation example:
  With:

inner diameter of the core	$2\,r_i = 250$ mm	
outer diameter of the core	$2\,r_a = 355$ mm	
height of the core	$h = 40$ mm	
number of turns	$N_S = 1374$	
frequency	$f = 50$ Hz	$\omega = 2\pi f = 100\pi$ 1/s
primary current	$I_P = 2500$ A	

$$i_P(t) = \sqrt{2} \cdot I_P \cdot \sin \omega t = \sqrt{2} \cdot 2500 \sin \omega t [\text{A}]$$

follows from Eq. (3.11):

$$U_0(t) = 2 \cdot 1374 \cdot 0.04 \cdot \ln\frac{\frac{355}{2}}{\frac{250}{2}} \cdot 10^{-7}\frac{\text{Vs}}{\text{A}} \cdot \frac{\partial}{\partial t}\left(\sqrt{2} \cdot 2500 \cdot \sin \omega t\right)\text{A} \qquad (3.14)$$

$$U_0(t) = 2 \cdot 1374 \cdot 0.04 \cdot \ln\frac{\frac{355}{2}}{\frac{250}{2}} \cdot 10^{-7}\text{Vs} \cdot \omega\left(\sqrt{2} \cdot 2500 \cdot \cos \omega t\right) \qquad (3.15)$$

and with $\omega = 314$ 1/s:

$$U_0(t) = 4.28 \cdot \cos \omega t [\text{V}] \qquad (3.16)$$

3.4.2   The Voltage U_S at Burden Z_b of the Air Core Coil

In the air core coil, the secondary windings are applied to a non-magnetic core. The air core coil has no saturation flux density B_{max} and has no hysteresis. The output variable is a linear function of the primary current under quasi steady state conditions.

From the equivalent circuit diagram in Fig. 3.66, equations are derived which show the influence of the circuit variables and the connected burden Z_B on the output variable and the secondary voltage $U_S(t)$. To simplify the calculation, the cable capacitance C_c is not considered in the derivations of the equations.

The primary current induces the voltage $U_0(t) = M \cdot \frac{\partial I_P(t)}{\partial t}$ in the coil. For the quasi-steady state and the sinusoidal primary current, the induced voltage in the air-core coil is $U_0 = M \cdot j\omega \cdot I_P$.

From the schematic diagram in Fig. 3.66 the secondary voltage is calculated for the connected burden Z_B:

$$U_s = \frac{Z_B}{R_S + Z_B + j\omega L_S} \cdot U_0 = \frac{Z_B}{R_S + Z_B + j\omega L_S} \cdot M \cdot j\omega I_P$$

For a large burden resistance $R_B = |Z_B| \to \infty$ is

$$U_S = U_0 = j\omega M \cdot I_P \text{ or } I_P = -j\frac{U_S}{\omega M}$$

The positive phase angle between the secondary voltage U_S and the primary current I_P is +90°.

3.4.3 The Phasor Diagram of the Air Core Coil

From the equation for the secondary voltage U_S follows the equation by transformation:
$I_P = \frac{U_S}{\omega M} + \frac{R_S}{Z_B} \cdot \frac{U_S}{\omega M} + j\frac{\omega L}{Z_B} \cdot \frac{U_S}{\omega M}$ for the phasor diagram in Fig. 3.68

From the phasor diagram in Fig. 3.68 follows the amplitude error ε for the air core coil:

$$\varepsilon = \frac{\frac{U_S}{\omega M} - I_P}{I_P} = \frac{U_S/\omega M}{\sqrt{\left\{\frac{U_S}{\omega M} + \frac{R_S}{Z_B} \cdot \frac{U_S}{\omega M}\right\}^2 + \left(\frac{U_S}{\omega M}\right)^2 \cdot \left(\frac{\omega L}{Z_B}\right)^2}} - 1$$

$$= \frac{1}{\sqrt{\left(1 + \frac{R_S}{Z_B}\right)^2 + \left(\frac{\omega L}{Z_B}\right)^2}} - 1 \approx \frac{1}{\sqrt{1 + \frac{2R_s}{Z_B}}} - 1 \approx -\frac{R_S}{Z_B}$$

with $Z_B \gg \omega L$ and $Z_B \gg R_S$.

A growing burden resistance reduces the amplitude error.

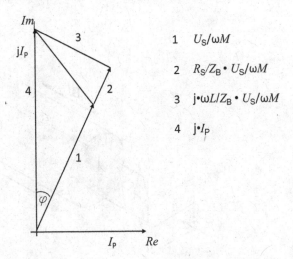

Fig. 3.68 The phasor diagram of the air core coil

1   $U_S/\omega M$

2   $R_S/Z_B \cdot U_S/\omega M$

3   $j \cdot \omega L/Z_B \cdot U_S/\omega M$

4   $j \cdot I_P$

The phase error $\Delta\varphi$ *in* relation to the phase shift of $+90°$ between the output variable U_s and the input variable I_p results to

$$\tan\Delta\varphi = \frac{\omega L_S/Z_b \cdot U_S/\omega M}{\frac{U_S}{\omega M} + \frac{R_S}{Z_B} \cdot \frac{U_S}{\omega M}} = -\frac{\omega L}{Z_B + R_S} \qquad \Rightarrow \qquad \Delta\varphi = -\frac{\omega L}{Z_B + R_S}$$

The negative phase error of the secondary voltage U_s and the $+90°$ leading primary current I_p grows with the inductance and is reduced with increasing values of $Z_B + R_S$.

3.5   Optical Current Transformers (Faraday Effect)

As an alternative to the inductive current transformer principle described in the previous chapters, the current can also be measured optically using the Faraday effect.

3.5.1   The Principle of Optical Current Measurement—The Faraday Effect

As early as 1845, Michael Faraday discovered that the polarisation direction of a linearly polarised light wave rotates when passing through an isotropic medium under the influence of a magnetic field in the direction of the light beam (Fig. 3.69). The angle of rotation β is proportional to the magnetic field H and the path length l within the material.

$$\beta = V \cdot H \cdot l$$

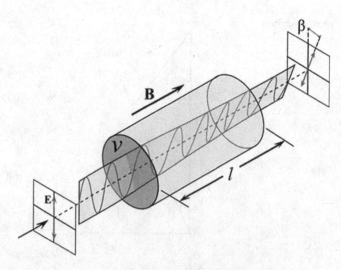

Fig. 3.69 Rotation of the polarisation direction due to the Faraday effect

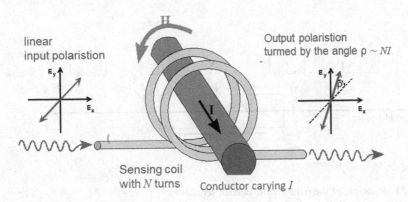

Fig. 3.70 Optical current measurement principle

The proportionality constant V, the Verdet constant, depends on the material and the wavelength of the light.

This magneto-optical effect is used to determine the current in a conductor by placing the optical medium (glass fibre or glass ring of prisms) around the current-carrying conductor. By applying the Ampere's Law $I = \oint H dl$, the current can be determined directly from the rotation of the polarisation direction (Fig. 3.70).

Physically, the Faraday effect can be explained as follows: The refractive index in the optical medium is changed by an applied magnetic field, depending on the polarisation state of the light. The refractive index for right-hand circularly polarised light and for left-hand circularly polarised light is different. This is called circular birefringence.

Linearly polarised light can be regarded as a superposition of right-hand circularly polarised and left-hand circularly polarised light. If the light runs in an optical medium under the influence of a magnetic field, the right-circularly and left-circularly polarised waves have a phase difference due to the different speeds. When superimposed, this phase difference causes a linear polarisation with a rotated polarisation direction.

In a polarimetric current sensor, the rotation of the linear polarisation is measured by means of a polariser and an analyser.

In contrast, the interferometric current sensor measures the phase shift of right and left circularly polarised light.

3.5.2   Polarimetric Current Sensor

With the polarimetric current sensor, the rotation of the polarisation direction is determined by means of two polarisers. The first polariser generates a linearly polarised light beam, which is then passed through the sensor element. A second polariser, called analyser, after the sensor element then allows only a part of the light to pass through, corresponding to the rotation of the direction of polarisation, and thus the angular rotation and thus the current can be determined by measuring the intensity of the light (Fig. 3.71).

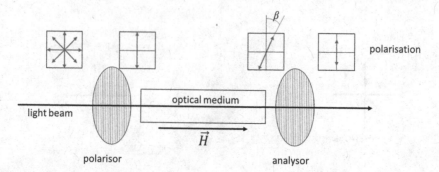

Fig. 3.71 Principle of the polarimetric current sensor

Fig. 3.72 Sensitivity of
current measurement by
rotating the analyser by 45°

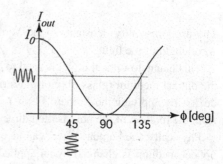

The intensity of the light after passing through the polariser and analyser is

$$I_2 = I_1 \cdot cos^2\beta$$

Since the angular rotations of the current sensors are small and the sensitivity of the cos^2 function is small around the zero point, the analyser is rotated by 45° with respect to the polariser, so that the operating point is shifted into the sensitive range of the cos^2 function, as shown in Fig. 3.72.

To avoid fluctuations in the intensity of light, the intensity I_2 at the output of the sensor is normalised to the intensity I_1 at the input of the sensor. This then results in the current in the conductor:

$$I = V \cdot \beta = \arccos\left(\frac{I_2 - I_1}{I_1}\right)$$

3.5.3 Interferometric Current Sensor (Sagnac Sensor)

In the interferometric current sensor, two circularly polarised light waves are passed through the sensor and at the end the phase shift of the two waves is measured and evaluated. Interferometers based on the Sagnac principle are used for this purpose, as used in gyroscopes for measuring the angle of rotation in missiles. The structure (see Fig. 3.73) is somewhat more complex than the polarimetric principle [50].

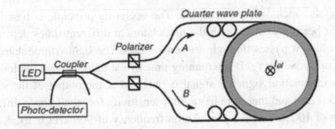

Fig. 3.73 Basic structure of an interferometric current sensor

The laser light is split into two beams in a beam splitter. Each beam is linearly polarised and then circularly polarised by two $\lambda/4$ elements. At the entrance of the sensor loop there are two circularly polarised light waves which propagate in opposite directions through the loop. Without an electric current both waves "see" the same refractive index. They therefore need the same time to pass through the loop and interfere constructively at the beam splitter. When a current flows through the conductor, the magnetic field caused by it points in the direction of propagation of one light wave and opposite the direction of propagation of the other. The two waves no longer need the same time to pass through the loop. Therefore, they are phase shifted by an angle $\Delta\varphi$ at the point of interference. The current can be determined by measuring this phase difference.

The interferometric sensor, as described above, has the disadvantage of being insensitive to small phase differences. The signal of the detector is dependent on $\cos(\Delta\varphi)$. This disadvantage can be corrected by inserting an optical phase modulator between beam splitter and sensor loop (Fig. 3.74).

In practice, this is done using a commercially available integrated optical circuit originally developed for use in optical gyroscopes. This device uses the electro-optical properties of lithium niobate to phase modulate light. The functions of the beam splitter and

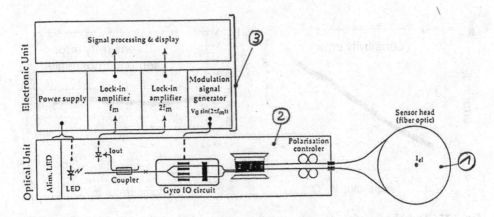

Fig. 3.74 Block diagram of an optical current sensor with Sagnac interferometer [50]

polarisers are also included in the device. The operating principle is based on the fact that each light beam passes through the modulators at different times depending on the direction in which it passes through the sensor loop. The light is modulated by a sinusoidal signal of frequency f_m. If the running time of the light through the loop is half the period of the modulation signal, a signal is obtained at the output of the sensor which depends on $\sin(\Delta\varphi)$ and therefore has a high sensitivity for small phase differences. For a sensor loop of 100 m length a modulation frequency of 900 kHz is used. An analysis of the signals at f_m and $2f_m$ allows to obtain a signal of the current which is independent from possible fluctuations of the light intensity of the source. [50]

3.5.4 Features of the Optical Current Sensor

Temperature and vibration are two parameters which can influence the output signal of the optical current sensor. This is one of the reasons why IEC standardisation has introduced tests for the sensitivity of current measurements to vibration and the influence of temperature on the measured signal [35].

The Verdet constant V and thus the angular rotation of the polarisation in the magnetic field depends on the temperature. This disturbing effect must be compensated for in the evaluation of the signal. There are different possibilities for this. A common solution is to measure the temperature at the sensor head and thus correct the signal in the processing electronics. Figure 3.75 shows measured values of an optical current transformer with and without temperature compensation.

Mechanical loads on the sensor elements due to vibration or bending of the glass fibre (especially with sensors made of glass fibre windings) lead to disturbing birefringence in the current sensor. This must be kept as low as possible to achieve the current measurement accuracy required for current transformers.

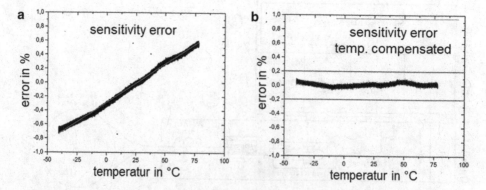

Fig. 3.75 Temperature influence on the optical current transformer **a** Measurements without temperature compensation **b** Measurements with temperature compensation

3.5.5 Design of Optical Current Transformers

An optical current transformer consists of 3 main parts.

a) Optical sensor head
b) Electronic processing unit
c) Current transformer interface (analogue or digital)

The Optical Sensor Head
Either glass fibre coils or glass rings made of glass prisms are used as the sensor element of the optical current transformer (see Fig. 3.76).

The Electronic Processing Unit
Each optical current transformer requires an electronic processing unit. This contains the light source for the sensor including the corresponding control electronics, photocells for measuring the light coming back from the sensor and the processing of the signals to determine the measured current.

The Interface of the Optical Current Transformer
Since analogue interfaces of conventional measuring transformers have high output powers, they are hardly suitable for optical current transformers. Analogue interfaces with low output powers, as used by low-power current transformers as described in Sect. 3.3 and by Rogowski coils as described in Sect. 3.4, however, can also be realised for optical current transformers.

Since the electronic processing of the signals in the optical current transformer is almost always made digital, the new digital interface for instrument transformers

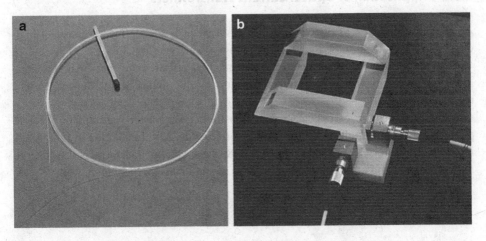

Fig. 3.76 Examples of optical sensing elements **a** Glass fibre coil, **b** Glass ring

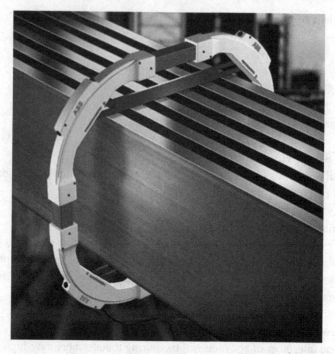

Fig. 3.77 Example of an optical sensor for measuring high direct currents (ABB) [51]

according to the standards IEC 61850 and IEC 61896-9 is best suited for the optical current transformers.

3.5.6 Application of Optical Current Transformers

A frequent application of optical current measurement is for the measurement of high direct currents, such as those found in aluminium production (Fig. 3.77).

With these direct currents up to several hundred amperes are measured. Previous measuring systems using Hall effect sensors can be replaced by optical current measurement.

For the application of optical current transformers in power transmission, several products have been available on the market for several years. In addition to free-standing optical transformers as shown in Fig. 3.78, these current transformers can also be mounted directly on other components, such as circuit breakers. (see Fig. 3.79).

Fig. 3.78 Example of a free-standing optical current transformer (Trench Germany) [52]

Fig. 3.79 Optical current transformer directly connected to a circuit-breaker (ENEL Terna substation Candia/Italy)

Voltage Measurement 4

4.1 Introduction and Standardisation

Until a few years ago, only conventional voltage transformers were used for voltage measurement. Inductive voltage transformers work according to the principle of the transformer. Their characteristics, requirements and designs are described in Sect. 4.2.

Another conventional voltage transformer is the capacitive voltage transformer. This consists of a capacitive divider, which reduces the high voltage to be measured to a voltage value of 10 to 20 kV, and an inductive intermediate transformer, which provides the output voltage. The capacitive voltage transformer is described in Sect. 4.3.

With the development of measuring and protection devices as digital electronic systems, the high output powers of voltage transformers are no longer needed. For this purpose, new non-conventional voltage transformers have been developed over the last 20 years which transform the primary voltages with high accuracy but no longer transmit power. The voltage divider technology is used for this purpose. For use in medium voltage, ohmic dividers are used, as described in Sect. 4.7, and for use in high voltage, mainly RC dividers, as described in Sect. 4.8. The necessary technology of resistors and capacitors are described in Sect. 4.4 and 4.5.

In addition, there are also optical voltage transformers which measure the voltage according to the Pockels effect. This is described in Sect. 4.9.

Conventional voltage transformers are only suitable for a limited frequency response, whereas RC dividers are suitable for a wide range from 0 Hz up to a few kHz, to transmit the primary signals accurately (see Sect. 4.6).

The requirements for voltage transformers are described in the IEC standard series IEC 61869. Part IEC 61869-3 [53] describes the requirements for inductive voltage transformers, IEC 61869-5 [54] describes the requirements for capacitive voltage transformers. The voltage dividers belong to the category of low-power instrument transformers, whose general requirements are described in IEC 61869-6 [35]. The special requirements for R-dividers and RC-dividers without electronics are in IEC 61869-11 [55] and the requirements for electronic voltage transformers in IEC 61869-7 [56]. The latter include optical voltage transformers and RC dividers with associated amplifiers.

Another part of the series of standards, IEC 61869-15 [57], describes the requirements of voltage transformers for use in DC systems.

4.2   Inductive Voltage Transformers

4.2.1 Principle of the Inductive Voltage Transformer

A conventional inductive voltage transformer is in principle a transformer with an open secondary winding as shown in Fig. 4.1.

The secondary voltage U_S is proportional to the primary voltage U_P and the ratio of the number of turns of the secondary winding N_S and the primary winding N_P. The secondary voltage can be calculated to $U_S = U_P \cdot N_S / N_P$. The accuracy of the voltage transformer depends on the size of the iron core, the resistance of the primary winding, the leakage inductance, and the connected burden.

Fig. 4.1 Principle of the
inductive voltage transformer

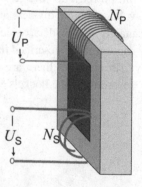

4.2.2 Errors of the Inductive Voltage Transformer and Their Influencing Factors

4.2.2.1 The Equivalent Circuit Diagram of the Voltage Transformer

Figure 4.2 shows the equivalent circuit diagram of the inductive voltage transformer, reduced to the secondary side. The primary winding is represented by the ohmic resistance Rp of the primary winding and the leakage inductance L_p, the secondary winding by the ohmic resistance R_S of the secondary winding and the secondary leakage inductance L_S. The behaviour of the iron core is represented by the main inductance L_m and the ohmic losses of the core R_{Fe}. The impedance of the external burden is represented by the ohmic part R_B and the inductive part L_B. The primary values are converted to the secondary side. Here applies:

$$U'_P = U_P N_S/N_P = U_P \cdot \ddot{u}, \quad I'_P = I_P N_S/N_P = I_P/\ddot{u}$$
$$R'_P = R_P (N_S/N_P)^2 = R_P \cdot \ddot{u}^2, \quad L'_P = L_P (N_S/N_P)^2 = L_P \cdot \ddot{u}^2,$$

With the transmission ratio $\ddot{u} = N_S/N_P$

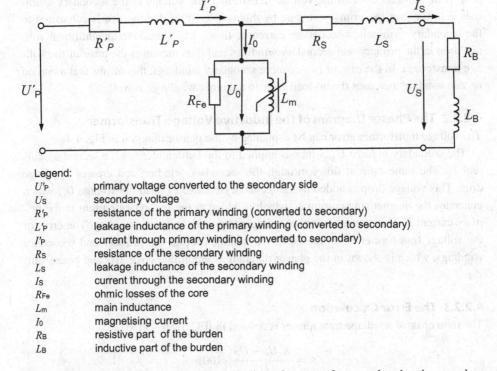

Legend:

U'_P	primary voltage converted to the secondary side
U_S	secondary voltage
R'_P	resistance of the primary winding (converted to secondary)
L'_P	leakage inductance of the primary winding (converted to secondary)
I'_P	current through primary winding (converted to secondary)
R_S	resistance of the secondary winding
L_S	leakage inductance of the secondary winding
I_S	current through the secondary winding
R_{Fe}	ohmic losses of the core
L_m	main inductance
I_0	magnetising current
R_B	resistive part of the burden
L_B	inductive part of the burden

Fig. 4.2 Equivalent circuit diagram of an inductive voltage transformer reduced to the secondary side

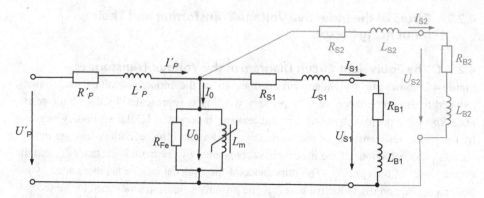

Fig. 4.3 Equivalent circuit diagram of a voltage transformer with 2 secondary windings

If there is more than one secondary winding in the voltage transformer, the equivalent circuit diagram must be extended for accuracy considerations, as shown in Fig. 4.3. Each secondary winding is shown with its winding and its burden.

If no burden is connected to the secondary terminals, the magnetizing current I_0 flows in the primary winding. This leads to a voltage drop at the primary winding. This voltage drop is the no-load error of the voltage transformer, the voltage at the secondary terminals is lower than the primary voltage by this voltage drop. If a burden is connected to the secondary terminals, a secondary current I_S flows, which leads to an additional voltage drop in the primary and secondary windings and thus increases the error of the voltage transformer. In the case of two or more secondary windings, this means that a burden on one winding increases the no-load error in the other windings as well.

4.2.2.2 The Phasor Diagram of the Inductive Voltage Transformer

The voltage transformer error can be explained by the phasor diagram in Fig. 4.4.

The secondary voltage U_S, which is applied to the burden, defines the secondary current I_S. The same current flows through the secondary winding and creates a voltage drop. This voltage drop is added to the secondary voltage to form the voltage U_0, which generates the magnetising current I_0, which is added to the secondary current to the primary current I_P. This creates a further voltage drop at the primary winding. The error of the voltage transformer is composed of the voltage drop at the primary and secondary windings, which is shown in the phasor diagram as amplitude error ΔU and phase shift $\Delta\varphi$.

4.2.2.3 The Error Calculation

The ratio error of a voltage transformer is defined in IEC 61869-3 [53] as:

$$\varepsilon = \frac{K_r U_s - U_P}{U_p} \cdot 100\%$$

Fig. 4.4 The phasor diagram
of the inductive voltage
transformer according to the
equivalent circuit diagram
Fig. 4.2 (reduced to the
secondary side)

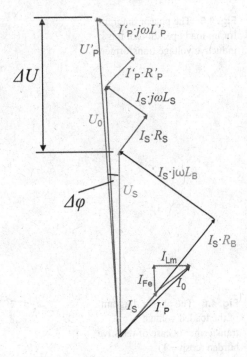

The rated transmission ratio K_r is the ratio of the rated value of the primary voltage U_{Pr} to the rated value of the secondary voltage U_{Sr}. Normally this is equal to the ratio of the primary to the secondary number of turns.

$$K_r = \frac{U_{Pr}}{U_{Sr}} = \frac{N_P}{N_S} = \ddot{u}$$

Calculation of the amplitude error of a voltage transformer:

A: No-load error of the voltage transformer

As long as no burden is connected to the secondary terminals, the magnetizing current I_0 flows in the primary winding. This leads to a voltage drop at the primary winding. This voltage drop represents the no-load error of the voltage transformer because the voltage at the secondary terminals is lower than the primary voltage by this voltage drop. In this case $U_S = U_0$ and $I_1 = I_0$ (Fig. 4.5).

B: Error with connected burden

If a burden is connected to the secondary terminals, a secondary current I_S flows, which leads to additional voltage drop in the primary and secondary winding and thus increases the error of the voltage transformer. Figure 4.6 shows the phasor diagram of the voltage transformer with ohmic burden ($\cos\beta = 1$). For complex burdens with ($\cos\beta < 1$) the primary voltage changes along the burden curve, which is also shown in Fig. 4.6. The burden curve is valid for a constant total burden $|Z_B| = \sqrt{R_B^2 + (\omega L_B^2)}$.

Fig. 4.5 The phasor diagram
for no-load operation of the
inductive voltage transformer

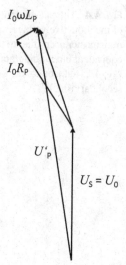

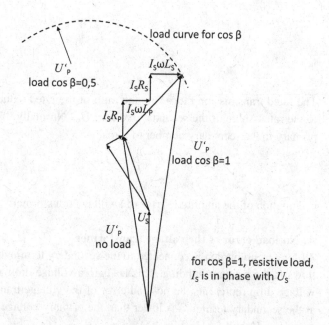

Fig. 4.6 The phasor diagram
of the loaded voltage
transformer in case of resistive
burden ($\cos\beta = 1$)

For the different burdens of an inductive voltage transformer the error diagram
according to Fig. 4.7 can be created from the above phasor diagrams. Here the amplitude
error ε and the phase shift $\Delta\varphi$ depending on the primary voltage (between 10% of the
rated voltage and 1.73 times the rated voltage) without burden (0 VA) and for a burden
of 300 VA at $\cos\beta = 0.8$ is shown. In addition, the values for $\cos\beta = 0.5$ and $\cos\beta = 1$ are
shown for the voltages 80% and 120% of the nominal voltage.

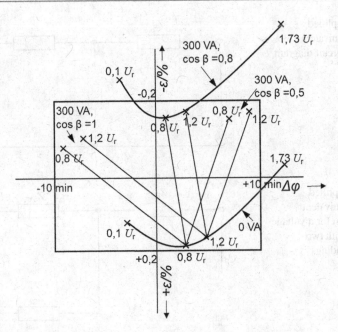

Fig. 4.7   The error diagram of an inductive voltage transformer for accuracy class 0.2

If necessary, the amplitude error can be shifted up or down by adjusting the number of primary windings.

C: Error with several loaded secondary windings

If there is more than one secondary winding in the voltage transformer, the additional burden must be considered when calculating the accuracy of a winding. According to the extended equivalent circuit diagram in Fig. 4.4 an additional voltage drop in the primary winding caused by the secondary current of the other secondary windings must be considered. The error of one winding therefore increases with the burden of the other winding.

Based on the simplified impedance equivalent circuit diagram of the inductive voltage transformer (see Fig. 4.8), the high main impedance Z_h can be neglected for the following evaluations.

The error voltage ΔU_S is determined by the voltage drop of the load current I_S at the primary (Z_P) and secondary impedance (Z_S). Therefore, the following applies:

$$\Delta U_S = I_S \cdot Z_P + I_S \cdot Z_S$$

For a voltage transformer with two secondary windings as shown in Fig. 4.9, the errors voltage of the two windings are calculated:

Fig. 4.8 Simplified impedance equivalent voltage transformer circuit diagram

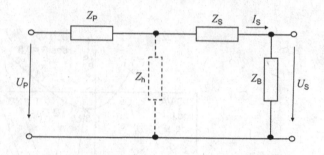

Fig. 4.9 Simplified impedance equivalent circuit diagram for a voltage transformer with two secondary windings

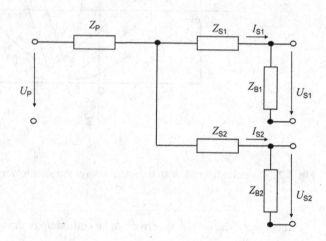

$$\Delta U_{S1} = I_{S1} \cdot Z_P + \boldsymbol{I_{S2}} \cdot \boldsymbol{Z_P} + I_{S1} \cdot Z_{S1}$$
$$\Delta U_{S2} = I_{S2} \cdot Z_P + \boldsymbol{I_{S1}} \cdot \boldsymbol{Z_P} + I_{S2} \cdot Z_{S2}$$

$\boldsymbol{I_{S2}} \cdot \boldsymbol{Z_P}$ and $\boldsymbol{I_{S1}} \cdot \boldsymbol{Z_P}$ are the shares of the error on one winding due to the burden on the other winding.

To avoid this influence, these additional errors must be compensated by a suitable circuit. A simple decoupling possibility for a transformer with two identical secondary windings is shown in the circuit in Fig. 4.10.

The result is now for the error voltages:

$$\Delta U_{S1} = I_{S1} \cdot Z_P + I_{S2} \cdot Z_P + I_{S1} \cdot Z_{S1} + \boldsymbol{I_{S1}} \cdot \boldsymbol{Z_E} - \boldsymbol{I_{S2}} \cdot \boldsymbol{Z_E}$$
$$\Delta U_{S2} = I_{S2} \cdot Z_P + I_{S1} \cdot Z_P + I_{S2} \cdot Z_{S2} + \boldsymbol{I_{S2}} \cdot \boldsymbol{Z_E} - \boldsymbol{I_{S1}} \cdot \boldsymbol{Z_E}$$

For complete decoupling, therefore, Z_E must be equal to Z_P:

$$\Delta U_{S1} = I_{S1} \cdot Z_P + I_{S1} \cdot Z_S + I_{S1} \cdot Z_P$$
$$\Delta U_{S2} = I_{S2} \cdot Z_P + I_{S2} \cdot Z_{S2} + I_{S2} \cdot Z_P$$

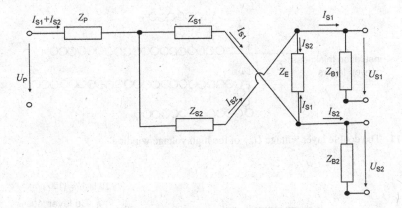

Fig. 4.10 Simple decoupling of two identical secondary windings

It can be seen that the error voltages are now independent of the burden on the other winding. However, the error component of the primary impedance Z_P increases, in the above shown symmetrical case it is doubled.

The decoupling impedance Z_E can also be coupled in a galvanic isolated manner via an additional intermediate transformer. In this way, even with different secondary windings, decoupling can be achieved by suitable selection of the number of windings of the intermediate transformer. Depending on the accuracy requirements of the windings, for example, the decoupling in one winding may be stronger than in the other winding.

4.2.3 Behaviour with Short-Circuited Secondary Winding

Inductive voltage transformers must not be operated with a short-circuited secondary winding. According to the requirements of the standards [53], however, they must withstand a 1 s short circuit of the secondary terminals without mechanical and thermal damage.

During normal operation of the voltage transformer, a relatively small current flows through the secondary winding. At a secondary voltage of 100 V and a burden of 100 VA, the secondary current is 1 A. However, if the secondary winding of the voltage transformer is short-circuited, high currents flow through the windings, which overheat the transformer. This then leads to the destruction of the voltage transformer.

4.2.4 Dielectric Stress on the Primary Winding

The double layer voltage is essential for dimensioning the layer insulation of the high voltage winding of the voltage transformer. Figure 4.11 shows the definition of the double layer voltage U_{DL} as the potential difference between the windings of two winding layers.

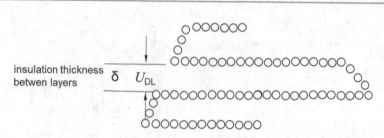

insulation thickness δ U_{DL}
betwen layers

Fig. 4.11 The double layer voltage U_{DL} of the high voltage winding

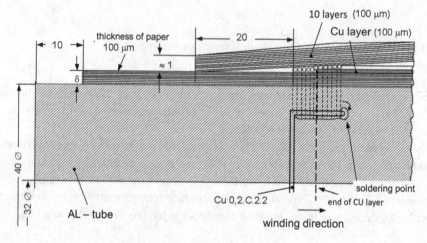

Fig. 4.12 Model arrangement to determine the partial discharge inception on a double layer

To dimension the primary winding of the voltage transformer, models for the double-layer voltage were produced as shown in Fig. 4.12. The partial discharge inception voltage was measured on these models.

For each model with different insulation thickness 5 pieces were produced and the partial discharge inception voltage was measured. The series of measurements were evaluated with the Weibull statistics and the voltage was determined, which leads to a partial discharge with a probability of 0.1% or 1%. This determines the allowed double layer voltage as a function of the insulation thickness, as shown in Fig. 4.13.

4.2.5   Single Stage Voltage Transformer

Today, inductive voltage transformers up to a voltage of 300 kV are mostly realised in single-stage design, as shown in Fig. 4.14. The iron core with the windings is located inside the housing which is at earth potential. The insulator can be a porcelain insulator, or a composite insulator made of FRP tube and silicone sheds.

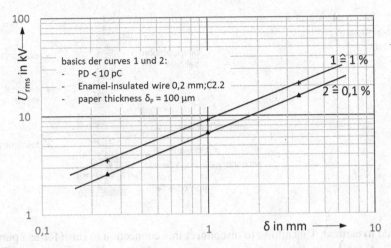

basics der curves 1 und 2:
- PD < 10 pC
- Enamel-insulated wire 0,2 mm;C2.2
- paper thickness δ_p = 100 µm

$1 \cong 1\%$

$2 \cong 0,1\%$

U_{rms} in kV

δ in mm

Fig. 4.13 Double layer voltage as a function of insulation thickness δ curve 1–1% probability for PD inception curve 2–0.1% probability for PD inception

Fig. 4.14 Single stage
inductive voltage transformers
for 123 kV

The primary winding and the secondary winding are applied to one leg of the iron core. The design of a voltage transformer for high-voltage application is shown in Fig. 4.15. The high voltage is connected through the bushing to the beginning of the high voltage winding. The end of the high voltage winding is led into the terminal box and

Fig. 4.15 Sectional view of a single stage inductive voltage transformer

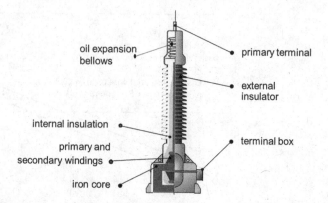

connected to earthed. This allows to disconnect this connection to earth for test purposes. However, this connection must always be guaranteed during operation.

In the single-stage inductive voltage transformer, the secondary winding is wound directly on the core and the primary winding on top of it. A high-voltage electrode is placed above the primary winding to control the electric fields in the insulation. To achieve better decoupling of any transient voltages that may occur on the primary side, a shield is inserted between the primary and secondary winding and connected to earth (Fig. 4.16).

The high-voltage winding is designed as a layer winding with insulation between the layers according to Sect. 4.2.4. The bushing is manufactured analogous to the current transformer bushing in Sect. 2.2 and has a capacitive potential grading made of aluminium foils.

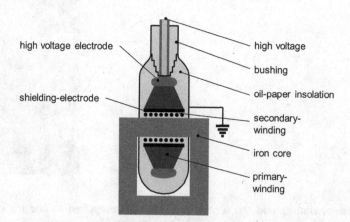

Fig. 4.16 The structure of the active part of the single stage inductive voltage transformer

4.2.6 Cascade Voltage Transformers

4.2.6.1 The Design of the Cascade Voltage Transformer

The primary winding of the magnetic voltage transformer and its insulation becomes disproportionately complex with increasing voltage. Therefore, for voltages above 300 kV, the inductive voltage transformer is in most cases designed as cascade design.

Figure 4.17 shows the principle of a cascade voltage transformer. The iron core and the housing are at half voltage potential.

The primary winding of the voltage transformer is divided into 2 halves and each half is insulated for half of the primary voltage. Both halves of the primary winding are wound around a common iron core which is at half the high voltage potential and is placed in a separate housing with the same potential.

The high voltage end of the upper half of the primary winding is connected to high voltage via a bushing, the earth end of the lower half of the primary winding is connected to earth via a second bushing.

The secondary winding is wound over the lower half of the primary winding and is also connected to the terminals in the terminal box via the lower bushing.

To achieve a balanced magnetic load on the core and to minimise leakage flux and thus voltage transformer errors, additional compensation windings are required on both legs of the magnetic core [1, 58]. This is described in detail in the following chapter.

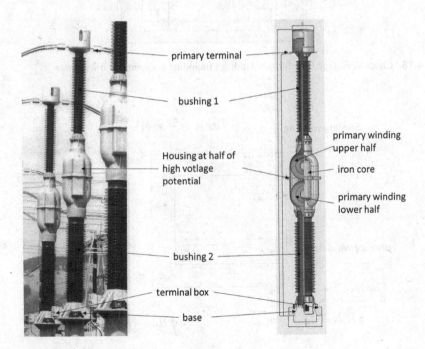

Fig. 4.17 Voltage transformer in cascade design

4.2.6.2 The Compensation Winding of the Cascade Voltage Transformer

Figure 4.18 shows the basic winding structure of a voltage transformer in cascade design.

Without the compensation winding, leakage fluxes occur due to the asymmetrical distribution of the windings, as shown in Fig. 4.19. On the left side the primary winding is drawn, divided into 2 halves with $N_P/2$ turns each and on the right side the secondary winding with N_S turns.

Figure 4.20 shows the compensation winding in the middle. This cancels out the flux outside the core and the leakage flux disappears.

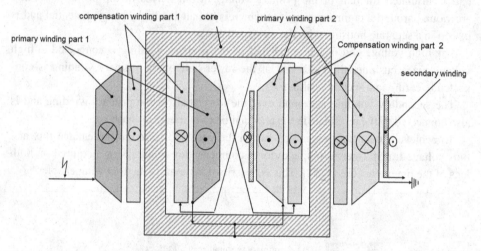

Fig. 4.18 Cascade voltage transformer windings including compensation windings

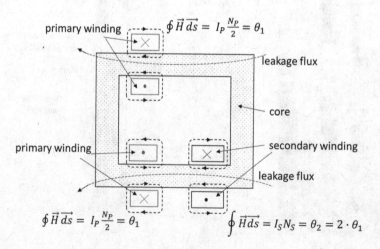

Fig. 4.19 Current linkages of a cascade voltage transformer without compensation winding

Fig. 4.20 Resulting excitation of a cascade design with compensation winding

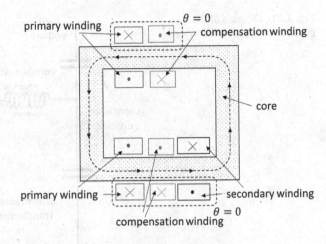

4.2.7 Ferro-Resonance Oscillations with Inductive Voltage Transformers

During switching operations in substations, steady-state, non-linear ferroresonance oscillations can occur.

Ferroresonance oscillations are oscillations between the non-linear inductances of the voltage transformer and the capacitances in the switchgear. Important parameters here are the grading capacitances of circuit breakers and the capacitances to earth. In the book "Ferroresonance Oscillations in Substations with inductive voltage transformers in medium and high voltage systems" [59] these ferroresonance oscillations are discussed in detail. There the theory, the occurrence of single-phase and three-phase ferroresonance oscillations, their simulation and detection are described. Additionally, remedial measures to avoid ferroresonance oscillations are shown.

4.3 Capacitive Voltage Transformers

4.3.1 The Theory of the Capacitive Voltage Transformer

Worldwide, 90% of the voltage measurements in high-voltage outdoor switchgear are made with capacitive voltage transformers (CVT). This is not the case for some of the utilities in Europe, which prefer the magnetic voltage transformer due to its reliability in accuracy, which has the advantage that the transformer either works correctly or is defective (e.g., due to a short circuit of the windings).

With the CVT, individual elements in the capacitor can break down without affecting the operational safety. However, the accuracy then no longer meets the requirements.

Figure 4.21 shows the typical circuit diagram of the capacitive voltage transformer.

Fig. 4.21 Diagram of the
capacitive voltage transformer

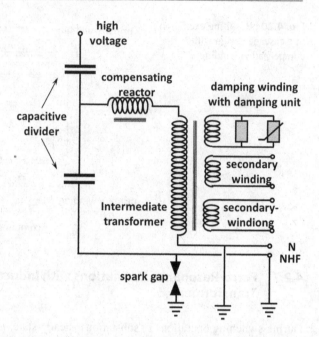

The CVT is divided into a capacitive divider and an electromagnetic unit with an intermediate transformer and a compensating reactor. The capacitive divider divides the primary voltage into an intermediate voltage of typically 10–15 kV. The intermediate transformer then converts this intermediate voltage into the output voltage at the secondary terminals. An additional damping winding is used to connect damping units to suppress ferroresonance oscillations (see Sect. 4.3.3).

From the schematic diagram in Fig. 4.21 follows the equivalent circuit diagram of the CVT in Fig. 4.22, where only one secondary winding is considered. The secondary side is converted to the primary side.

The output voltage of the voltage divider U_{C2}, which is applied to the compensating reactor and intermediate transformer, is calculated to:

$$U_{C2} = \frac{C_1}{C_1 + C_2} \cdot U_P$$

To dimension the compensating coil, the equivalent circuit diagram in Fig. 4.22 is transformed into the equivalent circuit diagram in Fig. 4.23 according to the theorem of Thevenin [60].

The input voltage is calculated as follows: $U_{ThP} = \frac{C_1}{C_1 + C_2} \cdot U_P$.

For dimensioning the compensating reactor, the reactances, and resistances are summarised as follows (see Fig. 4.24)

$$X_i = \omega L = X_{Dr} + X_{TP} + X_{TS} \cdot \ddot{u}^2$$
$$R_i = R_{Dr} + R_{TP} + R_{TS} \cdot \ddot{u}^2$$

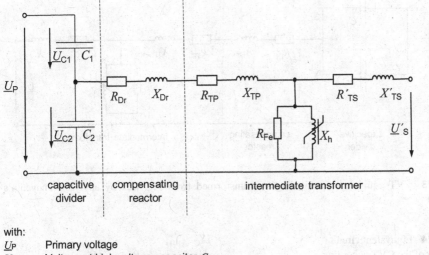

with:

U_P Primary voltage
U_{C1} Voltage at high voltage capacitor C_1
U_{C2} Voltage at intermediate voltage capacitor C_2
X_{Dr} Reactance of the compensating reactor
R_{Dr} Resistance of the compensating reactor
X_{TP} Primary leakage reactance of the intermediate transformer
R_{TP} Resistance of the primary winding
X_h Main reactance of the intermediate transformer
R_{Fe} Iron loss resistance of the intermediate transformer
X'_{TS} Secondary leakage reactance of the intermediate transformer (converted to primary side)
R'_{TS} Resistance of the secondary winding (converted on primary side)
U'_s Secondary voltage (converted to primary side)

Fig. 4.22 Equivalent circuit diagram of the capacitive voltage transformer

where $\ddot{u}$ is the transmission ratio of the intermediate transformer.

The compensating reactor must now be dimensioned so that the inductance is in resonance with the capacity of the divider. It must therefore apply:

$$\omega L - \frac{1}{\omega(C_1 + C_2)} = 0$$

To understand how the capacitive voltage transformer works, it is necessary to consider both the theorem of Thevenin and the standard equivalent circuit diagram.

Figure 4.25 shows the phasor diagram of the equivalent circuit in Fig. 4.24:

The example in Fig. 4.25 is not in resonance. The phasor $\underline{U}_{Xi}$ is too long and the phasor $\underline{U}_C$ too short. Resonance is present when $\underline{U}_{Xi} = \underline{U}_C$.

Figure 4.26 shows the standard equivalent circuit diagram of the CVT and Fig. 4.27 the corresponding phasor diagram. It cannot be seen from this diagram that the CVT is based on resonance matching.

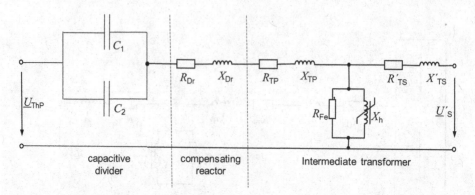

Fig. 4.23 CVT equivalent circuit diagram transformed from Fig. 4.22 by applying Thevenin's theorem

Fig. 4.24 Equivalent circuit diagram of the CVT according to the theorem of Thevenin under resistive-inductive load. (The main inductance and iron losses were neglected) where:
$-jX_C = \frac{1}{j\omega(C_1+C_2)}$

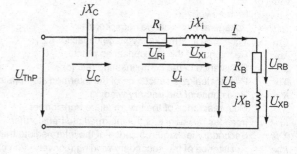

Fig. 4.25 Associated phasor diagram to Fig. 4.24

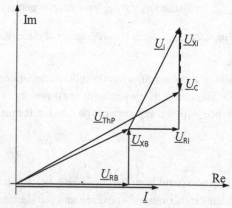

4.3.2 The Capacitive Voltage Transformer in Operation

The requirements on the capacitive voltage transformer depend on the tasks in the switchgear.

Figure 4.28 shows a capacitive voltage transformer in a 420 kV substation. The measuring system consists of a capacitive divider (in the case of the 420 kV CVT consisting

Fig. 4.26 Standard equivalent circuit diagram with resistive-inductive burden

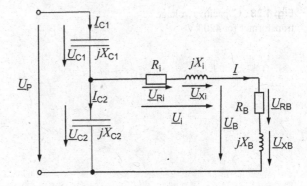

Fig. 4.27 Phasor diagram associated with Fig. 4.26

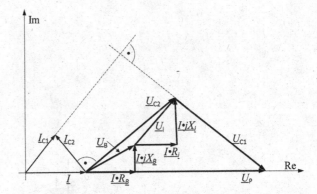

of 2 coupling capacitors), a compensating reactor and an inductive intermediate transformer.

The coupling capacitor mounted directly on the oil-filled electromagnetic unit has a tap which is electrically connected to the compensating reactor in the magnetic unit.

A sectional view of a capacitive voltage transformer is shown in Fig. 4.29.

For the application of the CVT as a protective voltage transformer, knowledge of the transfer function, response on the secondary side in case of a square-wave impact on the primary side, is required for planning the protection system.

IEC 61869-5 [54] specifies requirements for ferroresonance oscillations and the transfer function when short-circuiting the primary side,

Solving the problem of trapped charges, which is not mentioned in the IEC 61869-5 standard, is particularly difficult. It occurs when the CVT is switched off. If the voltage is not switched off at zero, the primary capacity of the voltage divider cannot discharge ("trapped charges" occur), but the secondary capacity discharges via the intermediate transformer. If the CVT is then switched on again, transient transition functions occur on the secondary side.

Figure 4.30 explains the problem of trapped charges.

Fig. 4.28 Capacitive voltage transformer for 420 kV

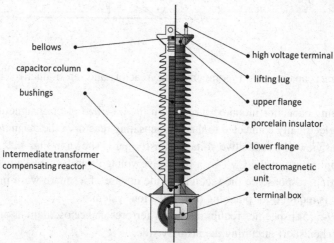

bellows

capacitor column

bushings

intermediate transformer
compensating reactor

high voltage terminal

lifting lug

upper flange

porcelain insulator

lower flange

electromagnetic
unit

terminal box

Fig. 4.29 Sectional view of a capacitive voltage transformer

When reconnecting the CVT, the voltage on the secondary side $U_{C2}(t)$ does not follow the primary voltage $U_1(t)$. A transient compensation process occurs which can lead to false tripping of protection relays.

Fig. 4.30 Trapped charges on the capacitive voltage transformer **a.** The measuring circuit of the CVT **b.** Voltage curve on the primary side of the dividers **c.** Voltage curve on the secondary side of the divider

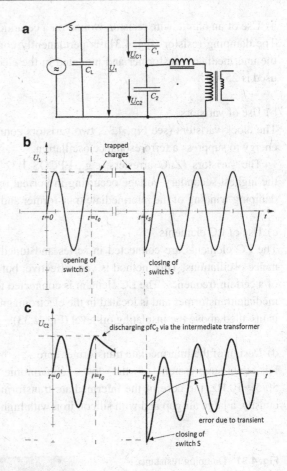

4.3.3 Ferroresonance Oscillations on the Capacitive Voltage Transformer

The CVT consists of a capacitive divider, a compensating reactor, and an intermediate transformer. These are elements that enable ferroresonance oscillation.

Excitation can be achieved by switching operations, such as switching on the CVT, or by shorting and re-opening the secondary terminals of the CVT.

The mathematical analysis of the ferroresonance oscillations shows that only odd subharmonics can occur, so for a 50 Hz system 50/1 = 50 Hz, 50/3 = 16.66 Hz and 50/5 = 10 Hz.

In practice, 4 methods for the suppression of ferroresonance oscillations have proven to be effective, which are listed below. The damping devices a) to c) are connected to an additional damping winding of the intermediate transformer (see Fig. 4.21).

a) Use of an ohmic wire resistor, wound on a ceramic tube.

The damping resistor (Fig. 4.31) is permanently connected to the damping winding of the intermediate transformer and mounted in the electromagnetic unit. A standard value used is 254 Ω.

b) Use of varistors

The block varistors (see Fig. 4.32, two varistors connected in series) can absorb enough energy to suppress a ferroresonance oscillation.

The varistors (ZnO arresters, e.g., EPCOS B32 K130/K1503) must not operate at the highest secondary voltage occurring in normal operation. They are connected to the damping winding of the intermediate transformer and installed in the terminal box.

c) Use of LC elements

The LC elements are connected in series and tuned to the frequency of the ferroresonance oscillations. This method is very effective, but it can only suppress the oscillation of a certain frequency. The LC element is connected to the damping winding of the intermediate transformer and is located in the electromagnetic unit. The metal paper capacitor is installed above the insulating oil level (Fig. 4.33).

d) Design of the intermediate transformer core

For the suppression of the low subharmonic ferroresonance oscillation with $50/5 = 10$ Hz, the core of the intermediate transformer must be designed with low flux density, a large air gap and with silicon iron with higher losses.

Fig. 4.31 Damping resistance in the electromagnetic unit

Fig. 4.32 Ferroresonance suppression varistors installed in the terminal box

Fig. 4.33 LC element (outlined) for ferroresonance suppression built into the electromagnetic unit of the CVT

4.3.4   The System for Carrier Frequency Transmission on High-Voltage Lines (PLC)

The PLC system (PLC—power line carrier) is still installed on some high-voltage lines to transmit messages. Coupling capacitors are used to feed in the signals. The capacitive divider of a CVT can also be used to couple the signal. As shown in Fig. 4.34, the signals are fed in between the earth end of the divider and earth.

4.3.5   The Main Elements of the Capacitive Voltage Transformer

Figure 4.34 shows a diagram of the capacitive voltage transformer with all its components, as well as the supply device for PLC signals.

The main elements in Fig. 4.34 are:

1. Voltage divider
2. Compensating reactor
3. Intermediate transformer

Further elements are:

4. Resistor and
5. Varistor for suppression of ferroresonance oscillation
6. Protection spark gap
7. Earth connection, which is removed when a PLC system is connected,
8. voltage limiter to protect the PLC system,
9. drain coil for the power frequency alternating current through the voltage divider.

Fig. 4.34 Diagram of capacitive voltage transformer with TFH system connected

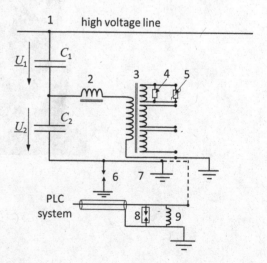

In the following subchapters the main components of the capacitive voltage transformer are described in more detail.

4.3.5.1 The Capacitive Voltage Divider

The capacitive voltage divider divides the primary voltage into an intermediate voltage in the range of 10–15 kV. It consists of capacitor elements with mixed dielectric and is built into the porcelain insulator. Figure 4.35 shows a sectional view of the divider with porcelain insulator and bushing.

Figure 4.36 shows the capacitor C_1 of the divider. It consists of 40 to 120 individual capacitor elements connected in series. The capacitor is composed of up to 2 packs, each of which is kept under constant pressure by means of compression springs to keep the capacitance constant.

The technology of the individual capacitor elements is described in Sect. 4.5.1.

The stacked capacitor elements are impregnated with insulating oil. To compensate the thermal expansion of the oil, oil expansion bellows are necessary, as shown in Fig. 4.35. Figure 4.37 shows a bellows used for oil expansion. It is filled with nitrogen at 1.1 bar overpressure and has an expansion range of $\pm 40\%$. Thus, a temperature range of $-50\,°C$ to $+70\,°C$ can be covered.

Oil-filled pressure cans are also used as alternative oil expansion bellows.

Two bushings are required for the connection of the divider tap to the compensating reactor and for the earth end of the divider. One possibility is to realise them in an epoxy resin part as shown in Fig. 4.38.

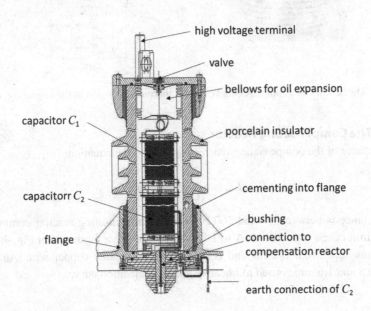

Fig. 4.35 CVT capacitive voltage divider

Fig. 4.36 Active part of capacitor C_1 of the voltage divider

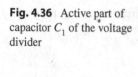

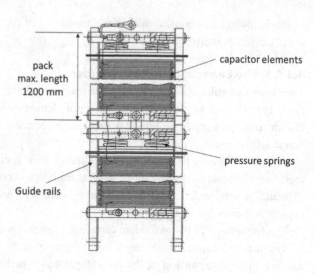

pack
max. length
1200 mm

capacitor elements

pressure springs

Guide rails

a   b

valve

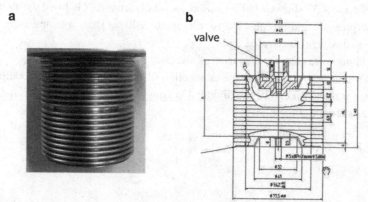

Fig. 4.37 The bellows as a compensating vessel for thermal expansion of the insulating oil

4.3.5.2 The Compensating Reactor

The inductance of the compensating reactor is given by the equation:

$$L_{Dr} = \frac{1}{(C_1 + C_2)\omega^2}$$

The inductance is between 50 and 500 H. For the compensating reactor, commercially available iron cores, e.g., SU 114A to DIN 41309 are used, as shown in Fig. 4.39. The winding has about 10,000 turns and is wound with insulated copper wire. An air gap between 0.8 and 1.6 mm is used to linearise the magnetisation curve.

Connection of the divider tap to the
compensating reactor

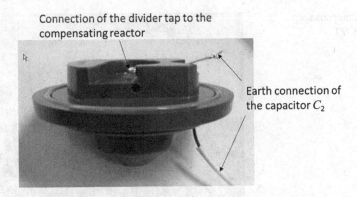
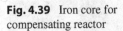

Earth connection of
the capacitor C_2

Fig. 4.38 Epoxy sealing of the divider with the bushings for the divider tap and the earth end of the divider

Fig. 4.39 Iron core for compensating reactor

Requirements for the compensating reactor:

- The reactor must block any signals from the PLC system,
- The reactor must be designed for the AC test voltage and lightning impulse voltage to IEC 61869,
- Easy adjustment of the resonance with the $C_1 + C_2$.

Figure 4.40 shows a finished compensating reactor. The core and winding must be isolated to the grounded steel tank for the voltage $U_{C2} = \frac{C_1}{C_1+C_2} \cdot U_P$.

For the adjustment of the compensating reactor to resonance with the capacitive divider, taps are attached, which can be adjusted when testing the CVT. Figure 4.41 shows these tapping.

Fig. 4.40 Compensating
reactor of a CVT

Fig. 4.41 Taps for
compensating reactor
adjustment

4.3.5.3 The Intermediate Transformer

The intermediate transformer transforms the voltage divided by the capacitive divider
into the secondary voltage. One or more secondary windings can be realised. In addition to the secondary windings, a damping winding is also applied in order to be able to
attach damping devices for ferroresonance oscillations (see sub Sect. 4.3.3).

Figure 4.42 shows schematically the structure of the intermediate transformer. The
primary winding is designed as a trapezoidal layered winding. Between the primary
winding and the rectangular secondary windings there is a shielding to avoid transient
interference signals on the secondary side. The shielding and the core are connected to
earth.

Fig. 4.42 Structure of the intermediate transformer

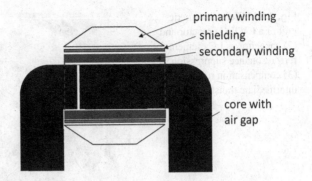

primary winding
shielding
secondary winding

core with
air gap

Fig. 4.43 The finished intermediate transformer with taps on the primary side

The primary winding is provided with taps on the high voltage side, which can be used to adjust the secondary voltage during the test (see Fig. 4.43).

4.3.5.4 The Electromagnetic Unit

The intermediate transformer and the compensating reactor are installed in a steel tank and impregnated with mineral oil. This electromagnetic unit together with the capacitive divider forms the capacitive voltage transformer. Figure 4.44 shows an electromagnetic unit with an LC element for ferroresonance suppression, consisting of a reactor (1) and a capacitor (2), installed in addition to the compensating reactor (3) and intermediate transformer (4). The capacitor is manufactured in metal-paper technology. This self-healing technology, in which metal is vapour-deposited on the paper dielectric, is described in detail in [61].

Fig. 4.44 Electromagnetic
unit of a CVT, (1) reactor and
(2) capacitor as LC element for
ferroresonance suppression,
(3) compensation reactor, (4)
intermediate transformer

4.4   Resistor Technology

4.4.1   Introduction to the Field of Voltage Measurement with Resistive Dividers

The first experiments in voltage measurement with resistive dividers were carried out in the field of high DC voltages for feeding electron and ion accelerators between 1955 and 1975 [62].

Depending on the application as actual value transmitter for a control loop, the measuring dividers had to have a high degree of temporal constancy [63].

For dividers used for electron accelerators in electron-microscopes stabilities of $f = \frac{u_i(t) - u_N(t)}{u_N(t)} \approx 2 \cdot 10^{-6}$ to $2 \cdot 10^{-5}$ have been reached.

In practice, two resistive divider technologies were used. The compensated R divider (see Sect. 4.7) used for voltages from 1 to 52 kV_{rms} AC and for DC, with compensation of earth capacitances by design measures. The RC divider (see Sect. 4.8) is used for AC voltages from 52 to 1200 kV_{rms} and for DC voltages up to 4000 kV, with resistors and capacitors connected in parallel.

Two metrological requirements for the use of compensated R dividers and RC dividers:

1. Parallel connection of capacitors
   Since the available measuring current is limited by the cost of the resistors and by the heating of the dividers at 200 µA to 600 µA with an operating voltage of 200 kV to 4 MV, the stray capacitances [64] force the use of capacitors connected in parallel.

2. Proportional transmission from the primary voltage $U_p(t)$ to the secondary voltage $U_s(t)$ of the measuring system

This requirement is necessary to achieve correct results for protective systems. It is not feasible that e.g., the problem of trapped charges in GIS or open-air systems can lead to correct signals on the secondary side of a C-divider and in capacitive voltage transformers during switching operations. [65] [66]

The industrial introduction of the compensated R-divider and the RC-divider in high voltage measurement technology for AC voltage was the lecture in Spokane in the USA [64]. The compensated resistor divider version measures frequencies from 0 Hz to 10 kHz, the RC divider version with parallel connected capacitors up to 1 MHz.

4.4.2 The Technology of Resistors

The essential influencing factors are the resistors connected in series for the measurement. Some necessary information for the selection of the resistors is given in Table 4.1.

The article [47] gives the main results. The use of metal film resistors has failed because rapid changes in the electrical fields and the associated forces destroy the metal film. The series-connected mass resistors have proven to be robust dividers but are unsuitable as precise actual value transmitters for control circuits for generating high stability of supplies. Wire resistors have been used in the RC divider of the electron accelerator [62].

Resistance values and resistance changes must not be measured with low voltage, but with operating voltage. The Schering bridge has proved to be a suitable measuring system (see Fig. 4.52).

The thermal noise of the resistors is recorded with an oscilloscope in Figs. 4.45, 4.46 and 4.47. The wire resistor has the lowest and the mass resistor the highest noise voltage. The thermal noise of different resistor technologies is recorded under the same measuring conditions in a bridge circuit with four 3 MΩ resistors [67, 68].

The noise voltage was measured at three different time scales. For the thick-film resistors used for cost reasons in recent years there are no noise measurement results available.

From the bridge circuit the noise voltage per resistor is calculated to be 8 times the measured peak-to-peak voltage.

Only the thick-film resistor is used as a replacement for the very costly wire resistors. In the ETZ article "Resistor technology for medium voltage sensors" [68], lifetime tests were carried out on high-voltage thick-film resistors (see Sect. 4.4.4).

Table 4.1 Necessary specifications of resistors in dividers

1	Temperature coefficient $T_K = f(T)$ [1/T] a) T_{Kmin}, b) $T_{Kmedium}$, c) T_{Kmax}
2	Type of resistors a) Wire Resistor b) Metal Film Resistor c) Thick film resistor d) mass resistor e) Cermet resistor
3	The resistors must be applied to full ceramic cylinders or plates, as flashovers can occur inside tubes
4	Maximum thermoelectric voltage between resistance material (e.g., manganin, resistance paste, carbon mass) and copper/tinned copper connections
5	Thermal noise of the resistors in the frequency range 5 kHz–1 MHz (see Figs. 4.45, 4.46 and 4.47)
6	Stability $\Delta R/R_0 = f$(years) at room temperature (initial value R_0 at 20 °C and 50% humidity
7	Level of internal temperature rise at maximum operating voltage
8	Thermal time constant
9	Maximum operating voltage
10	Dependence of the resistance value on the operating voltage $R = f(U_{rms})$
11	Long-term stability as a function of time and temperature $\Delta R/R_0 = f(t,T)$ in continuous operation
12	Change in resistance under stress with steep impulses and switching impulses

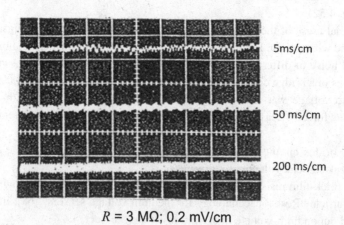

5ms/cm

50 ms/cm

200 ms/cm

$R = 3$ MΩ; 0.2 mV/cm

Fig. 4.45 Measured noise voltage on wire resistor $U_R \approx 8 \times 20$ µV $= 0.16$ mV

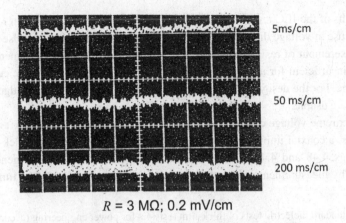

$R = 3 \text{ M}\Omega; \; 0.2 \text{ mV/cm}$

Fig. 4.46 Measured noise voltage at the cermet resistor $U_R \approx 8 \times 60 \; \mu V = 0.48 \text{ mV}$

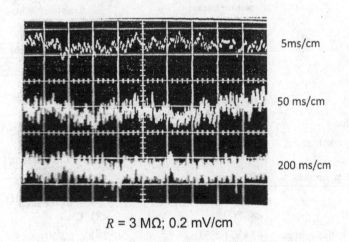

$R = 3 \text{ M}\Omega; \; 0.2 \text{ mV/cm}$

Fig. 4.47 Measured noise voltage at mass resistance $U_R \approx 8 \times 240 \; \mu V = 1.92 \text{ mV}$

4.4.3 The thick-film resistor

New test and measurement methods had to be developed for the application of thick-film resistor technology, because the type tests standardised in IEC 61869-1 and IEC 61869-11 were not sufficient to meet the requirements regarding the accuracy of the classes 0.2% or 0.1% and the long-term stability of the transformation ratio of the divider for metering application.

The change in resistance with temperature is very small. The temperature coefficient α_1 is between $10 \cdot 10^{-6} \text{ K}^{-1}$ and $50 \cdot 10^{-6} \text{ K}^{-1}$ depending on the temperature range.

The results of the IEC tests in Tables 4.2 and 4.3 show that the thick-film resistors are suitable for use in voltage dividers when exposed to AC and lightning impulse voltage.

The measurement of resistance values with electronic measuring instruments must be considered insufficient for accuracy measurement systems according to the current state of knowledge. For the design only resistors measured with the Schering bridge at operating voltage are usable.

For the extreme voltage changes with a front time of 5 ns to 10 ns during switching in GIS systems, a coaxial impulse generator was developed for testing the thick-film resistors (see Figs. 4.48 and 4.49). The self-inductance of the coaxial surge generator was determined by measurements using the resonance method to 1.5 µH. The time constant

Table 4.2 Standard dielectric tests of thick-film resistors for power engineering (according to IEC 61869-1 and -11)

Test parameters	Thick-film flat resistor	Thick-film round resistor
Nominal value	50 MΩ	35 MΩ
Temperature coefficient	α_1	α_1 (5 test samples) $\alpha_2 = \alpha_1/2$ (5 test samples)
Number of test samples	10	10
AC voltage	17.4 kV, 1 min, 50 Hz	11.5 kV, 1 min, 50 Hz
Temperature	<88 °C	<82 °C
Lightning impulse voltage	43 kV, 1.2/50 µs	28 kV, 1.2/50 µs
Number of impulses (per test specimen)	15 positive impulses 15 negative impulses	15 positive impulses 15 negative impulses

Table 4.3 Test results after exposure according to Table 4.2

Type	Dielectric behaviour	$\Delta R/R$ in % related to the resistance value before the tests
Flat resistor	No flashovers along the resistors	$+5 \cdot 10^{-3}$ to $-10 \cdot 10^{-3}$
Round resistor	No flashovers along the resistors	5 test samples (α_1) very small decrease in AC voltage and lightning impulse voltage 5 test items (α_2) Deviations not detectable with measuring method

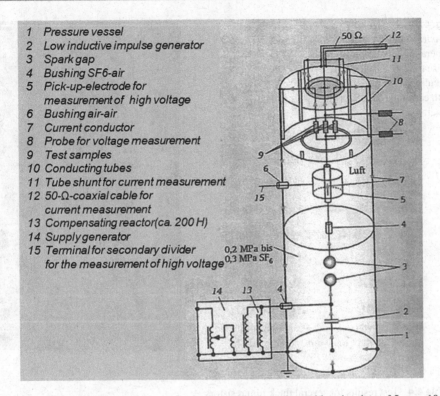

1 Pressure vessel
2 Low inductive impulse generator
3 Spark gap
4 Bushing SF6-air
5 Pick-up-electrode for
   measurement of high voltage
6 Bushing air-air
7 Current conductor
8 Probe for voltage measurement
9 Test samples
10 Conducting tubes
11 Tube shunt for current measurement
12 50-Ω-coaxial cable for
   current measurement
13 Compensating reactor (ca. 200 H)
14 Supply generator
15 Terminal for secondary divider
   for the measurement of high voltage

Fig. 4.48 Coaxial impulse generator for generating bipolar pulses with a rise time of 5 ns to 10 ns

$t = L/R$ of the system lies between $4 \cdot 10^{-5}$ ns and 0.1 ns, depending on the resistance value of the test specimens.

4.4.4 Lifetime testing of thick-film resistors with bipolar voltage impulses

In this test, round and flat resistors were measured which had passed the type tests according to IEC 61869-1 and 61869-11. The resistors were subjected to bipolar impulses with the coaxial impulse generator in Figs. 4.48 and 4.49.

The resistors were subjected to a large number of impulses and then the change in the resistance value was measured. Table 4.4 shows the results for different resistors.

The tested standard resistors show a resistance change of $\Delta R/R_0 \geq 1 \cdot 10^{-2}$ in case of steep impulse stress. These changes are too large for the application of resistive voltage dividers in the metering system.

The measurements of the value $\Delta R/R_0$ were made with a digital measuring instrument and not at operating voltage and are therefore not sufficiently meaningful.

The shape of the bipolar voltage impulses is shown in Fig. 4.50.

Fig. 4.49 Design of the
coaxial impulse generator for
generating bipolar pulses with
a rise time of 5 ns to 10 ns. 1
Impulse generator 2 Reactor
with earthing rod 3 Supply
generator

Table 4.4 Test results for several thick film resistors

Type	Number of test specimens	Data	Test voltage peak value (kV)	Number of impulses	*Result* $\Delta R/R$
Round resistors	5	$l=37$ mm, $d=8$ mm 35 MΩ, α_1	12	300,000	+0.7 to +0.9%
	5	$l=52$ mm, $d=8$ mm 35 MΩ, α_2	12	300,000	+0,9%
Flat resistors	1	$l=75$ mm, $b=20$ mm 50 MΩ, α_1	38	12,000	−7%
		$l=75$ mm, $b=20$ mm 50 MΩ, α_1	5 26 34 38	60,000 20,000 12,000 10,000	−1.6%

The further development of the thick-film resistors for use in high-voltage engineering gave better results for the same tests as in Table 4.4. The test results of the lifetime test of these improved resistors are given in Table 4.5.

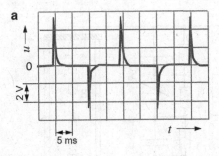 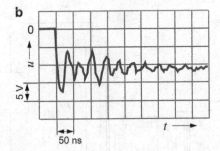

Fig. 4.50 Voltage impulses of the coaxial impulse generator **a.** bipolar voltage pulse **b.** single negative voltage pulse

Table 4.5 Test results with highly constant thick-film resistors

Quantity	Dimensions	Temperature coefficient	Resistance value (MΩ)	$\Delta R/R_0$
4	$l=75$ mm, $b=20$ mm	$\alpha_2=\alpha_1/2$	25	0 to –0.015%
1	$l=75$ mm, $b=20$ mm	$\alpha_2=\alpha_1/2$	25	–0.012%

The values $\Delta R/R_0$ were measured with digital instruments and were practically unrecordable. These resistances are suitable for the application of metering systems of the classes 0.1% and 0.2%.

For the evaluation of these thick-film resistors, the measurement with high voltage after repeated switching on and off of a test voltage with 36 kV$_{rms}$ is still missing.

4.4.5 The stress on the thick-film resistors by switching on and off the test voltage of 36 kV$_{rms}$

To check the influence of switching on and off the voltage on the thick-film resistors, a test arrangement was created which switches the test voltage of 36 kV on and off via a vacuum switch (see Fig. 4.51). In this way, 10,000 switching cycles (on and off) were made, and the resistance value was measured with a modified Schering bridge developed at Bern University in Burgdorf (see Fig. 4.52).

Tested were 8 series connected 20 MΩ resistors. So, the total resistance had a nominal value of 160 MΩ. The initial value before the switching operations was measured with the Schering bridge at 20 kV to $R_0 = 176.9017$ MΩ. The tests were carried out in the high-voltage laboratory of the Bern University of Applied Sciences in Burgdorf [69].

Table 4.6 shows the results of the resistance measurements. Three separate series of resistance measurements were carried out each with voltages of 5 kV, 10 kV, 15 kV, 20 kV and 25 kV with the Schering bridge as shown in Fig. 4.52 The mean value in Table 4.6 was used to calculate the deviation from the initial value.

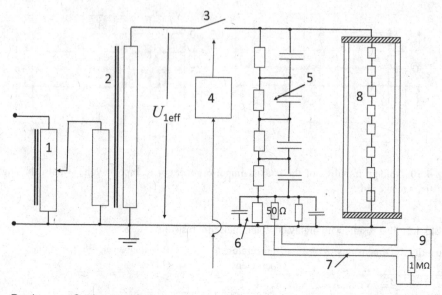

Pos 1   Setting transformer
Pos 2   High-voltage transformer
Pos 3   Vacuum switch
Pos 4   Signal transmitter for on/off switching
Pos 5   Reference voltage divider
Pos 6   coaxial secondary of reference divider
Pos 7   50 Ω Coaxial cable
Pos 8   Test sample
Pos 9   Oscilloscopes

Fig. 4.51 Test circuit for switching on and off the thick-film resistors

$$\frac{\Delta R}{R_0} = \frac{1/3(R_{\emptyset 1} + R_{\emptyset 2} + R_{\emptyset 3}) - R_0}{R_0} = \frac{R_\emptyset - R_0}{R_0}$$

Figure 4.54 shows three types of thick-film resistors. By connecting meandering resistor elements in series (see Fig. 4.54b), a very small inductance of the thick-film resistor is achieved, which is essential for a linear voltage distribution along the active part.

In the tests with switching on and off the voltage, resistors of the design shown in Fig. 4.54c were used with a nominal value of 20 MΩ.

Meandering elements are also connected in series on the round resistors shown in Fig. 4.54c. Here the distances between the tracks must be designed in such a way that there are no flashovers between the individual elements. The distance between the start and the conductor track of half the meander element must be designed with the appropriate flashover distance.

Measuring with the modified Schering bridge is a technically demanding engineering activity and requires knowledge of sensitive high-voltage technology. The balancing

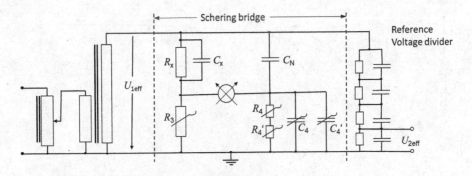

R_X	test sample, 8x 20 MΩ = 160 MΩ Thick-film resistors
C_X	additional capacity 20 pF (polystyrene capacitors)
R_3	10 kΩ
R_4+R_4'	100 Ω until 2000 Ω
C_N	974.4 pF, normal capacitor (compressed gas capacitor)
C_4	1.11 µF (bridge)
C_4'	19.48 µF additional capacitor (polystyrene)

Fig. 4.52 Modified Schering bridge for measuring high resistance

Table 4.6 Resistance measurements and deviation from the initial value of thick-film resistors under load with on/off switching operations

Number of switching operations	Measured resistance $R_\emptyset$ in MΩ	Deviation from initial value in %
100	176.9675	0.0372
500	177.5792	0.3830
1000	177.9539	0.5948
1500	178.1502	0.7058
2000	178.3925	0.8427
3000	178.4497	0.8751
4000	178.4961	0.9013
5000	178.5085	0.9083
6000	178.5698	0.9430
7000	178.5825	0.9501
8000	178.6030	0.9617
9000	178.6035	0.9620
10,000	178.6071	0.9640

of the bridge is not possible, for example, if partial discharges occur in one of the three high voltage components in the bridge R_x, C_x and C_N, or if the supply voltage contains harmonics.

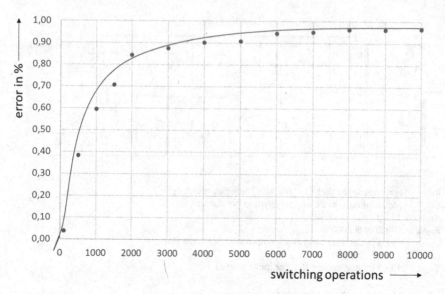

Fig. 4.53 Deviation of the resistance value from the initial value as a function of the number of switching cycles

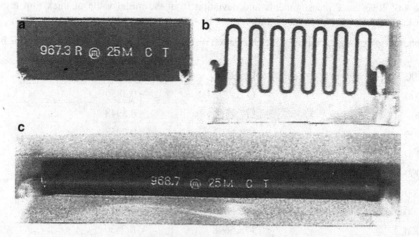

Fig. 4.54 Designs of thick film resistors **a**. Flat resistor on ceramic support covered with protective varnish, **b**. Flat resistor on ceramic plate without protective varnish. (7 meander elements connected in series), **c**. Round resistor on ceramic round bar covered with protective varnish

For the use of the voltage dividers in metering systems of accuracy classes 0.1 and 0.2, the value $\Delta R/R_0 \leq 0.05\%$ should not be increased by switching operations. According to Fig. 4.53, the resistors should therefore be pre-aged with approx. 5000 on/off cycles.

Fig. 4.55 Wireframe model of a 25 MΩ thick-film resistor to determine inductance

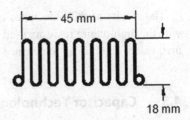

Other ways to improve the stability of the resistance value are the development of resistors with finer-grained pastes [70], or the development of another manufacturing process for the resistors.

In thick-film resistors which are installed in electric fields with high field strengths, the active part must be protected by a thin glass layer so that the field forces cannot tear particles out of the active part [47].

4.4.6 The Inductance of the Thick Film Resistors

The meandering resistor path, as shown in Fig. 4.54b, minimises the inductance of the resistance.

In a test the inductance was measured. For this purpose, a wire model of the resistor was produced, in which the wire was brought into the shape of the resistance path (see Fig. 4.55). The thickness of the wire was chosen so that its diameter (0.8 mm) corresponded to the width of the resistor path.

Table 4.7 shows the inductance and resistance measured at different frequencies. The model corresponds to a 25 MΩ thick-film resistor. Measurements were made using an "HP Precision RLC Meter 4284 A".

Table 4.7 Measured inductance of the resistor model from Fig. 4.55

Frequency	Inductance L (nH)	Resistance R (Ω)
500 Hz	50	8.80
1 kHz	170	8.80
2 kHz	210	8.80
5 kHz	216	8.80
10 kHz	207	8.80
20 kHz	202	8.80
50 kHz	198	8.80
100 kHz	195	8.80
200 kHz	192	8.80
500 kHz	188	8.81
1 MHz	186	8.83

The results of the inductance measurements show that inductance can be neglected with the high-ohmig resistors used in dividers. Even for a frequency of 1 MHz, the measured values result in an impedance of $Z_L = \omega L = 2 \cdot \pi \cdot 10^6 \ 1/s \cdot 186 \ 10^{-9} \ H = 1.17 \ \Omega$.

4.5   Capacitor Technology

4.5.1   Oil-Insulated Capacitors

4.5.1.1 The Oil-Paper Capacitor Element

Capacitors for use in high-voltage engineering are usually manufactured as a stack of individual wound elements.

The individual capacitor element is shown schematically in Fig. 4.56 and illustrated in Fig. 4.58. The capacitor elements are made with a mixed dielectric of polypropylene and high-density insulating paper.

When using this mixed dielectric, the temperature influence on the capacity is minimised. Insulating paper has a positive temperature coefficient, i.e., the capacitance of a capacitor with pure paper dielectric increases with rising temperature. In contrast, the temperature coefficient of polypropylene is negative, i.e., the capacitance of a capacitor with pure polypropylene dielectric decreases with increasing temperature. To obtain a capacitance that is as independent of the temperature as possible, a dielectric mixed from insulating paper and polypropylene is therefore used.

The capacitor elements are manufactured with a winding machine, as shown in Fig. 4.57, which is set up with different materials depending on the design of the elements.

The elements are stacked and pressed together by compression springs. To connect the individual elements, tinned copper lugs (see Fig. 4.58) are inserted and joined together in the stack (see Fig. 4.59).

Per connection 1 to 4 tinned copper lugs with a thickness of 60 μm and a width of 10 mm are used. The number of required lugs is determined by the energy stored in the capacitors.

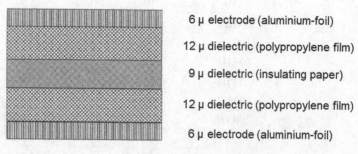

6 μ electrode (aluminium-foil)

12 μ dielectric (polypropylene film)

9 μ dielectric (insulating paper)

12 μ dielectric (polypropylene film)

6 μ electrode (aluminium-foil)

Fig. 4.56 Schematic structure of a capacitor element with mixed dielectric

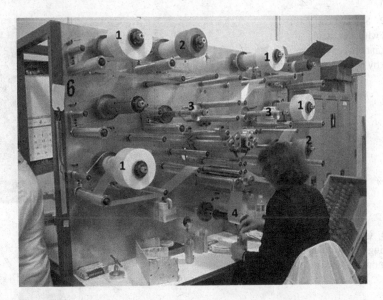

Fig. 4.57 Capacitor winding machine. 1 Polypropylene film, 2 insulating paper, 3 aluminum foil 4 wound capacitor element

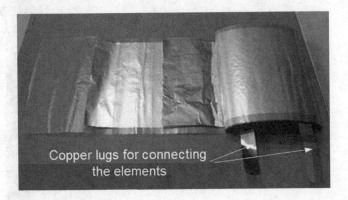

Fig. 4.58 Finished capacitor element

4.5.1.2 The Application of the Oil-Paper Capacitor

The stacks of capacitor elements are installed in porcelain or composite insulators and impregnated with oil. Besides mineral oil, synthetic oil such as Jarilec or PXE is also used for the capacitors.

The capacitors are used in capacitive voltage transformers (Sect. 4.3) and RC dividers (Sect. 4.8).

In addition, the capacitors are also used alone as coupling capacitors for coupling high-frequency signals of carrier-frequency signal transmission on high-voltage lines (PLC). The stable construction allows the coupling capacitor to be used also as a supporting structure for the high-frequency line traps necessary for the PLC system, as shown in Fig. 4.60.

Fig. 4.59 Stacked capacitor
elements with electrical
connection between the
elements by copper lugs and
pressure

Fig. 4.60 Coupling capacitor
for coupling PLC signals with
directly mounted line trap

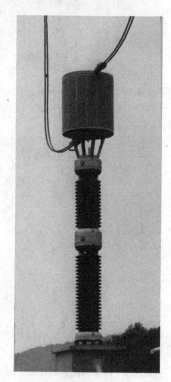

4.5.2   SF$_6$-insulated Capacitors

4.5.2.1 Factors Influencing the Capacity

In SF$_6$ insulated RC dividers as described in Sect. 4.8.4.2, capacitors with dielectric
made of polypropylene films are used.

For the use of gas-impregnated film capacitors for RC dividers, the change in capacitance as a function of temperature must be known.

Fig. 4.61 The difference in capacitance measured in percentage and the relative electricity constant for polypropylene as a function of temperature

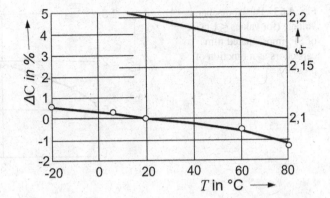

Figure 4.61 shows the dependence of the percentage capacitance difference $\Delta C = \frac{(C - C_{20})}{C_{20}} \cdot 100\%$ of the temperature. The dependence of the capacitance on the temperature in the range of $+20\,°C$ and $-20\,°C$ is low at $\pm 0.5\%$. In addition, the relative dielectric constant ε_r of polypropylene as a function of temperature is shown.

In addition to temperature, the influence of the applied voltage on the capacitance of the SF_6 impregnated film capacitors is also important. Figure 4.62 shows this influence. The results show that when the voltage is increased from $0.75\,U_N$ to $1.25\,U_N$, the capacitance increases by 0.14%, which is a small value.

The loss factor of the capacitor flat windings is difficult to measure because the losses of the guiding rails are parallel to the losses of the capacitors. The dissipation factor $\tan \delta$ of the polypropylene film is less than $0.5 \cdot 10^{-3}$.

4.5.2.2 The Lifetime of SF$_6$ Insulated Film Capacitors

The occurrence of partial discharge as a function of SF_6 pressure is an essential factor for the lifetime of polypropylene capacitors.

As shown in Fig. 4.63, an increase in pressure above 5 bar does not bring any improvement in partial discharge inception voltage. The pressure in gas-insulated switchgear is in many cases about 5–5.5 bar absolute.

Fig. 4.62 Capacitance of the SF$_6$ film capacitor as a function of the applied voltage

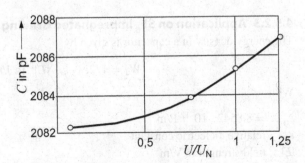

Fig. 4.63 Partial discharge voltage (for values ≤ 1 pC) of SF_6 impregnated film capacitors as a function of pressure

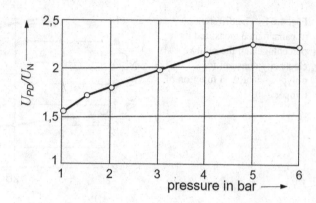

Fig. 4.64 Apparent charge in pC as a function of standardised voltage for different values of SF_6 pressure in bar

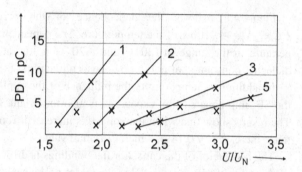

The graph in Fig. 4.64. shows the apparent charge of the partial discharges in pC as a function of the applied voltage for different SF_6 pressures. The higher the SF_6 pressure, the smaller the increase in apparent charge when the voltage is increased.

The determination of the lifetime curve in Fig. 4.65 was carried out with SF_6 film capacitors with a dielectric made of 20 μm polypropylene at 20 °C and a pressure of 5 bar absolute.

Increasing the pressure from 1.5 bar absolute to 5 bar absolute increases the lifetime of the capacitor by a factor of 4.

4.5.2.3 Application on SF_6 Impregnated Grading Capacitors

The energy density in a capacitor is given by:

$$W_d = 1/2\varepsilon_0 \cdot \varepsilon_r \cdot |E|^2 \text{ in J/m}^2$$

With

ε_0 $= 8.8542 \cdot 10^{-12}$ F/m
ε_r relative dielectric constant
$|E|$ field strength in V/m

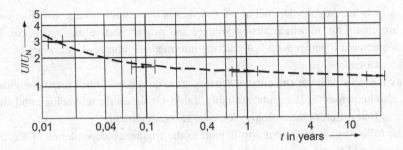

Fig. 4.65 Lifetime curve of SF$_6$ film capacitors

Fig. 4.66 Energy density as a function of field strength for capacitors with **a**) SF$_6$ films, and **b**) high-density paper impregnated with mineral oil as dielectric

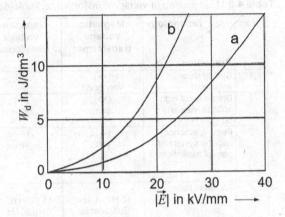

The calculation of the curves in Fig. 4.66 was performed with:

a) $\varepsilon_r = 2.1$ for polypropylene and
b) $\varepsilon_r = 5$ for oil-impregnated high-density insulating paper

Since the field strength of SF$_6$-impregnated polypropylene as dielectric can be selected 50% higher, the same energy density is achieved as with oil-impregnated paper capacitors. Since the bellows is omitted in the gas-filled capacitor, the volume of the capacitor is smaller.

4.6 Voltage Divider for Measuring High Voltages

Due to the centralisation of power generation, the phasing out of nuclear energy, the shutdown of coal and gas-fired power plants and the feeding of electrical energy from many small power plants with alternative energy from photovoltaic, wind and biogas plants, which are largely dependent on the weather, medium-voltage distribution networks are emerging, in which strong voltage and power fluctuations can occur. Local control equipment together with fast reacting intermediate storage is necessary to control these fluctuations.

Many universal voltage and current sensors are required to control these complex networks, also known as "smart grids", reliably. Table 4.8 shows the technology and characteristics of four different technologies for voltage measurement.

In the following, special attention is paid to the frequency dependence of the output voltage as an additional assessment of the various voltage measurement systems.

Table 4.8 Comparison of voltage transformer technologies [71]

No.	Technology / Properties	Magnetic voltage transformer	Capacitive voltage transformer	C-voltage divider	Compensated R- or RC- voltage divider
1	Steady state accuracy	(++) very good	(+) well	(+) well	(++) very good
2	Discharge of high-voltage lines and busbars	(++) yes	(- -) no	(- -) no	(+) yes, with large time constant
3	Ferro-resonance with the system or natural oscillation	(- -) yes	(- -) yes	(++) no ferroresonance oscillations cannot occur	(++) no ferroresonance oscillations cannot occur
4	Transient behaviour	(- -) 10 Hz...1 kHz Sufficient for slower protection functions	(- -) 45 (55) Hz...55 (65) Hz Additional damping or relay correction necessary	(+) 20 Hz...2 MHz Sufficient for the usual protection functions	(++) very good 0 Hz...2 MHz
5	Short circuit on the secondary side	(- -) Fuses necessary to prevent damage	(- -) Fuses necessary to prevent damage	(++) can be short-circuited for long periods without damage	(++) can be short-circuited for long periods without damage
6	Dielectric behaviour against impulse overvoltages	(+) external voltage distribution must be forced by grading. Impulse-resistant layer windings require high engineering effort.	(++) linear internal and external voltage distribution. External voltage distribution is given by the internal voltage distribution.	(++) linear internal and external voltage distribution. External voltage distribution is given by the internal voltage distribution.	(++) linear internal and external voltage distribution. External voltage distribution is given by the internal voltage distribution.
7	Burdens range	(++) Burden must be within specification	(-) Ferroresonance damping increases with burden	(-) Transmission ratio and phase displacement change with burden	(+) can be easily adjusted

Assessment: (- -) insufficient, (-) partly sufficient, (+) good, (++) very good

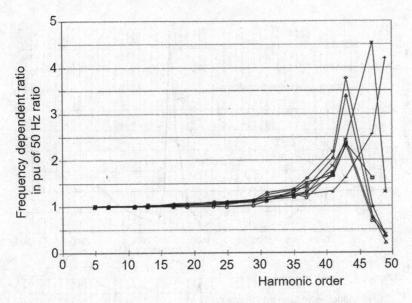

Fig. 4.67 Transformation ratio as a function of frequency comparing different 145 kV inductive voltage transformers [72]

Figure 4.67 shows measurements of the frequency response of various 145 kV inductive voltage transformers. All these voltage transformers have a resonance point at the 40th to 45th harmonic, i.e., at 2000 Hz to 2500 Hz. Larger errors of 10% are expected at a frequency of 1 kHz already.

Figure 4.68 compares the frequency behaviour of different voltage transformer technologies for use in 420 kV networks. The curves show the capacitive voltage transformer (see Sect. 4.3), which is tuned to the grid frequency, the inductive voltage transformer (see Sect. 4.2), which shows different resonance points, the first at about 400 Hz, and the RC voltage divider (see Sect. 4.8), which shows constant transmission characteristics in the frequency range shown up to 3 kHz. The RC voltage divider is therefore superior to the other technologies in terms of transmission of higher frequencies. Additionally, the RC voltage divider also measures the DC voltage. Inductive voltage transformers and capacitive voltage transformers cannot measure DC voltages.

Figure 4.69 shows the frequency behaviour of a pure ohmic R-divider without compensation (see Sect. 4.7). Existing stray capacitances to earth make the transformation ratio dependent on the location of the installation. In the measured arrangement the divider would only be usable up to a frequency of 10 Hz. To improve the behaviour of this resistive divider, the compensated R-dividers have been developed as described in Sect. 4.7. These compensated R-dividers are used as "low-power passive voltage transformers" according to IEC 61869-11 [55] in medium voltage systems up to 52 kV.

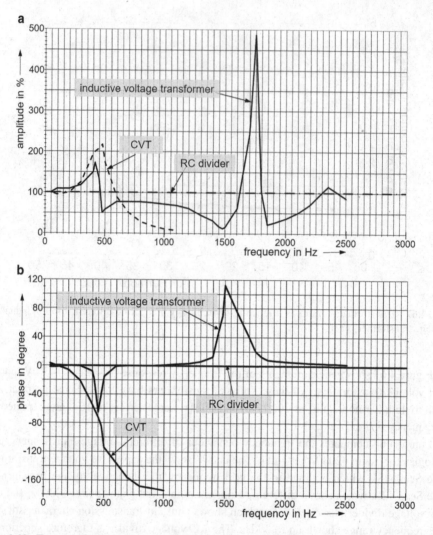

Fig. 4.68 Transmission behaviour of 420 kV voltage transformers of different technologies. **a**. Amplitude response as a function of frequency **b**. Phase response as a function of frequency

4.7   Compensated Resistive Dividers

4.7.1   The Theory of Compensated Resistive Dividers

In practice, two types of compensated R-dividers are produced:

a) The cylindrical/concentric arrangement shown in Fig. 4.70.

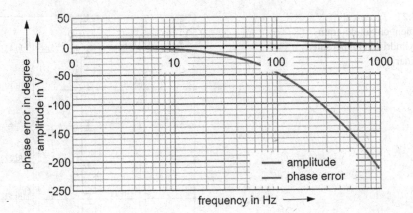

Fig. 4.69 Amplitudes and phase errors as a function of frequency of an R-divider without compensation

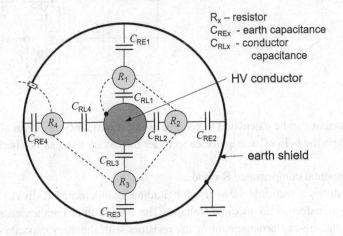

Fig. 4.70 Section through the basic structure of a cylindrical R-divider

Figure 4.70 of a compensated R-divider shows 4 round resistors R_1 to R_4 of a low-inductance design, printed with a meander-shaped resistor track. Ceramic low inductive flat resistors can also be used.

The corresponding electrical equivalent circuit diagram is shown in Fig. 4.71 as a kind of chain conductor.

The compensation of the amplitude response and the phase response over the frequency, i.e., the avoidance of the influence of the stray capacitances C_{RE} against earth, is achieved by the stray capacitances C_{RL} against the high voltage conductor.

In the equivalent circuit diagram Fig. 4.71 the high voltage resistors are framed with a dotted line and have a virtual tap in the middle. With the help of a computer program the

Fig. 4.71 Electrical equivalent circuit diagram of a cylindrical compensated R-divider

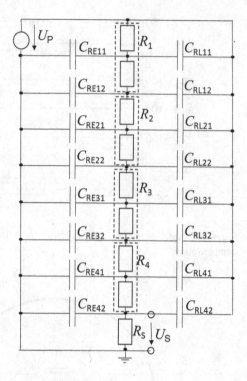

stray capacitances can be calculated to achieve an ideal frequency response of amplitude and phase. With the help of an impedance analyser the calculation can be checked.

b) The longitudinal compensated R-divider

Figure 4.72 shows a sectional view of a longitudinal compensated R-divider as applied as a voltage transformer in medium voltage. The longitudinal compensation is easily achieved by the zig-zag arrangement of the resistors with the stray capacitances C_{P1} to C_{Pn}.

The voltage divider is a high-impedance network and is protected against external electrical fields by the cylindrical metal tube pos. 1.

A further requirement for medium voltage sensors is insensitivity to external magnetic fields in the very compact medium voltage switchgear where high currents flow in the busbars. The inner network must therefore not have any loops in which voltages can be induced.

The equivalent circuit diagram in Fig. 4.73 is the basis for calculating the compensation of the earth capacitances C_{e1} to C_{en} by the longitudinal capacitances C_{P1} to C_{Pn}.

Figure 4.74 shows the equivalent electrical circuit of a serially produced compensated R-divider for $U_m = 24$ kV (Table 4.9). The connection to the secondary equipment is a double shielded cable with twisted wires. The outer screen is connected to earth in the divider housing.

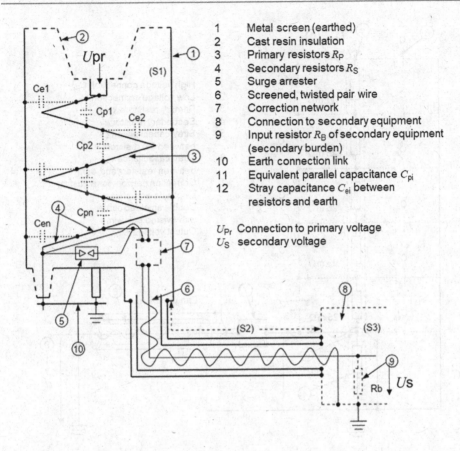

1	Metal screen (earthed)
2	Cast resin insulation
3	Primary resistors R_P
4	Secondary resistors R_S
5	Surge arrester
6	Screened, twisted pair wire
7	Correction network
8	Connection to secondary equipment
9	Input resistor R_B of secondary equipment (secondary burden)
10	Earth connection link
11	Equivalent parallel capacitance C_{pi}
12	Stray capacitance C_{ei} between resistors and earth

U_{Pr} Connection to primary voltage
U_S secondary voltage

Fig. 4.72 High voltage divider, shielded with longitudinal compensation C_{P1} to C_{Pn}

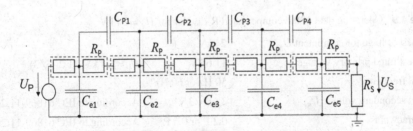

Fig. 4.73 Equivalent electrical diagram of a longitudinally compensated R-divider

The output power is greatly reduced compared to inductive voltage transformers. The use of R-dividers as voltage transformers requires a non-traditional voltage input circuit of the secondary technology, such as protection devices and meters. The input impedance must be as high as possible (>1 MΩ) for not influencing the accuracy.

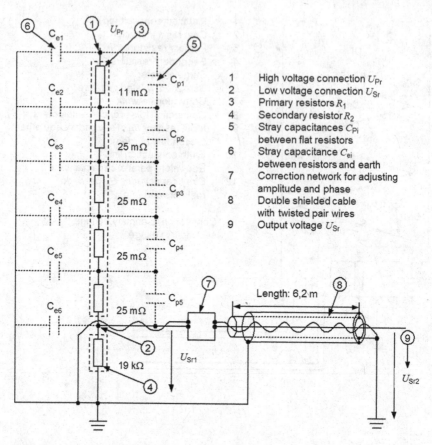

Fig. 4.74 Longitudinal compensated R-divider for $U_m = 24$ kV

Table 4.9 Operating data of the compensated R-divider for $U_m = 24$ kV

Highest voltage for equipment U_m	24 kV	
Highest rated primary voltage U_{Pr}	11.6 kV	$U_{Pr} \leq 24$ kV $/(1.2 \cdot \sqrt{3})$
Rated frequency	50 Hz or 60 Hz	
Rated secondary voltage U_{Sr}	$3.25/\sqrt{3}$ V	According to IEC 61869-11:2017
Accuracy class	0.2 P or 0.5 P	According to IEC 61869-11:2017
Power loss at continuous max. primary voltage	1.75 W	$P = (1.2 \cdot U_{Pr})^2/111 \cdot 10^6 \Omega$
Power loss at a voltage factor of 1.9 (8 h)	4.38 W	$P = (1,9 \cdot U_{Pr})^2/111 \cdot 10^6 \Omega$
AC voltage test 1 min	50 kV	
Lightning impulse test 1.2/50 μs	125 kV	
Current at rated voltage	100 μA	$I = U_{Pr}/111 \cdot 10^6 \Omega$

The R-dividers for medium voltage applications are insulated with cast resin. The dimensioning of the insulation is described in Sect. 1.4.

The insulation is designed according to the following 6 criteria:

1. AC voltage test
2. Impulse voltage test
3. Chopped impulse voltage test
4. External temperature range (day and night)
5. Service lifetime >40 years
6. PD inception voltage $\geq$ AC test voltage with value ≤ 2.5 pC

Criteria 1, 4 and 5 are closely related. There are components in the high voltage range which have excellent strengths for alternating voltage and surge voltages, but which fail after 10 years.

In practice, it has been shown that, depending on the connection technology between the resistors, the distance must be greatly increased to comply with the partial discharge condition (≤ 2.5 pC at AC test voltage).

A further influence on the insulation clearances are the parameters for the service lifetime >40 years at a continuous load of $1.2 \cdot U_{Pr}$. Particular attention must be paid to the large temperature differences between day and night (e.g., in Mexico, Brazil, Australia) in the expected service lifetime.

4.7.2 The Design of the Compensated R-Divider

4.7.2.1 The Cylindrical Compensation

Figure 4.75 shows the basic principle of cylindrical compensation, the spiral arrangement of the primary resistors R_{Pr} (1), the primary conductor (2) and the shielding (3) at earth potential. For rotationally symmetrical conductor design, the spiral arrangement of the resistors is the simplest solution for compensation.

Positive phase angles can also be generated by stray capacitances against the conductor (4) and against the shield (5).

The compensation is calculated with the electrical equivalent circuit diagram in Fig. 4.71 and the design structure in Fig. 4.75 using a suitable computer program.

Figure 4.76 shows the integration of a compensated R-divider into a combined instrument transformer, consisting of a compensated R-divider, low-power current transformer and medium-voltage bushing. Figure 4.77 shows this combined instrument transformer as a cast resin version for $U_m = 12$ kV.

Despite the combination of the three functions, a compact device is possible, which enables the design of cost-saving medium voltage systems.

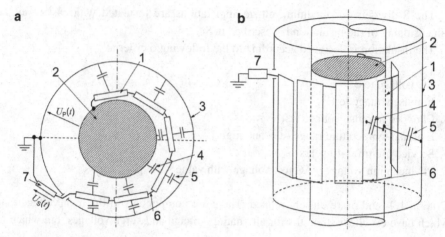

(1)   Primary resistance R_{Pr}
(2)   Primary conductor at high voltage potential
(3)   Shielding at earth potential
(4)   Stray capacitance between resistor and primary conductor
(5)   Stray capacitance between resistor and shielding
(6)   Connection between primary resistors
(7)   Secondary resistor R_S

Fig. 4.75 Schematic diagram of the compensated R-divider with spiral arrangement of the primary resistors. **a.** Section through the compensated R-divider **b.** perspective view of the compensated R-divider

4.7.2.2 The Longitudinal Compensation

Figures 4.72, 4.73 und 4.74 show the longitudinal compensation for R-dividers. The individual resistors are designed in a zig-zag pattern (see Fig. 4.78) to utilise the stray capacitance between the resistors to compensate for the influence of the stray capacitance against earth. The distances between the resistors are adjustable for optimum compensation.

Figure 4.79 shows another R-divider before casting with cast resin. It shows the resistor support (1), which is a spacer and defines the defined position of the primary resistors, which is needed for compensation. For a reproducible longitudinal compensation, small tolerances are necessary for the spacer. The high-voltage electrode (3) is at the same time connection to the high-voltage conductor in the switchgear, as well as shielding.

Figure 4.80 shows a cast resin insulated combined instrument transformer, consisting of a low-power current transformer with integrated shunt (see Sect. 3.3) and a longitudinal compensated R-divider.

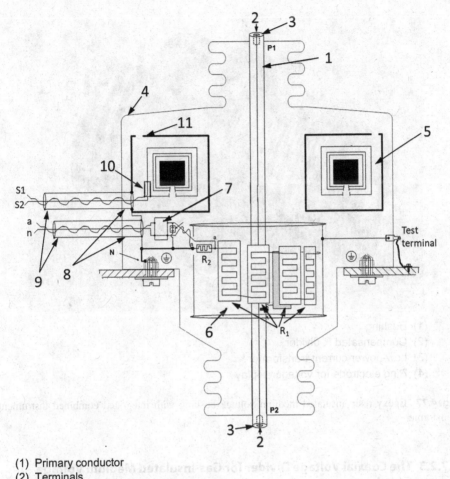

(1) Primary conductor
(2) Terminals
(3) Connecting surfaces of primary terminals
(4) Epoxy resin insulation
(5) Low-power current transformer with integrated shunt
(6) Compensated R-divider
(7) Correction network to the compensated R-divider
(8) Shielded twisted-pair cable inside the transformer
(9) 6 m shielded twisted-pair connection cable
(10)Shunt (integrated burden of the current transformer
(11)Current transformer shielding with separation

Fig. 4.76 Combined instrument transformer consisting of compensated R-divider, low-power current transformer and medium voltage bushing

(1) Bushing
(2) Compensated R-divider
(3) Low-power current transformer
(4) Ring electrode for voltage display

Fig. 4.77 Epoxy resin insulated medium voltage bushing with integrated combined instrument transformer

4.7.2.3 The Coaxial Voltage Divider for Gas-Insulated Medium Voltage Switchgear

The divider shown in Fig. 4.81 was developed for use in single-phase encapsulated gas-insulated medium-voltage switchgear [73].

The compensation of temperature and stray capacitance are not covered in this chapter, but the installation of the high voltage flat resistors is described.

Figures 4.82 and 4.83 show the influence of the installation position of the flat resistors on the potential lines within the divider. The horizontal installation according to Fig. 4.82 shows concentrations of potential lines and thus increases in field strength along the installation of the resistors. In contrast, Fig. 4.83 shows a practically ideal course of the potential lines due to the vertical installation of the resistors.

From the design point of view, the solution shown in Fig. 4.82 was very practical and easy to implement. However, measurement of the partial discharges during the test at 70 kV ($U_m = 36$ kV) and the requirement for a partial discharge value<2.5 pC showed that this design was useless.

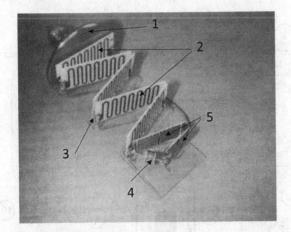

1. High voltage electrode
2. Low inductive flat resistors (meandering printed resistor track on ceramic plate
3. Connection between the resistors
4. Surge arrester
5. Secondary resistor

Fig. 4.78 Active part of a longitudinal compensated R-divider

Fig. 4.79 Assembled
longitudinal compensated
R-divider

1. Spacer for resistors
2. Low inductive flat resistors
3. High voltage electrode and shielding
4. Connection between the resistors

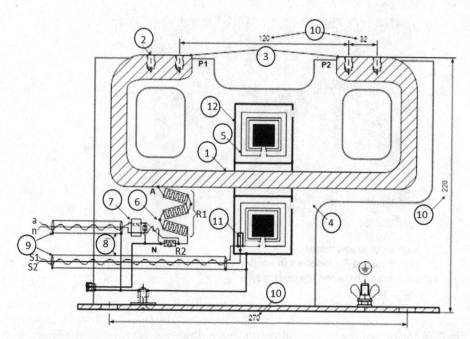

1. Conductor at high voltage potential
2. Primary terminal for the current transformer
3. Distance of the primary terminals
4. Cast resin body
5. Low-power current transformer with shunt and voltage output
6. Longitudinally compensated voltage divider
7. Correction network of the voltage divider
8. Shielded cable inside the combined instrument transformer
9. Output cable, shielded with twisted pair wires
10. Mounting dimensions
11. Shunt in low-power current transformer
12. Shielding of current transformer

Fig. 4.80 Sectional view of a cast resin combined instrument transformer with low-power current transformer and longitudinal-compensated R-divider

The solution shown in Fig. 4.83 with vertically installed flat resistors gave good partial discharge values <2.5 pC at the test voltage of 70 kV.

The calculation of the field strength vectors in Figs. 4.82b and 4.83b was based on the knowledge of the field strength for coaxial arrangements $\left|\vec{E}\right| = \frac{U_{rms}}{r \cdot \ln r_2/r_1} \left[\frac{kV}{mm}\right]$ which is not influenced by the vertical resistance chain.

With a diameter of the high voltage electrode of 132 mm and a diameter of the grounded flange of 226 mm, field strengths of 2.0 kV/mm at the high voltage electrode and 1.2 kV/mm at the inner side of the flange result.

The assessment of the field strengths in Figs. 4.82 and 4.83 does not show excessive field strengths at the fixing points of the resistors. At the test voltage of 70 kV, field

Fig. 4.81 Voltage divider for GIS medium voltage switchgear

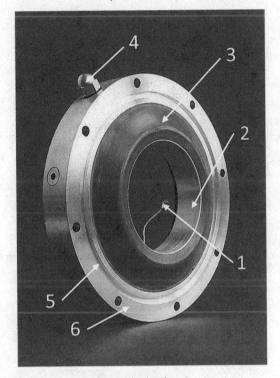

1. High voltage connection to the busbar
2. High voltage shielding electrode
3. Cast resin insulation (epoxy/PU resins)
4. Earth screw of the R-divider
5. O-ring groove
6. Aluminium flange

strengths of 3.7 kV/mm were determined for horizontal resistors and 5.0 kV/mm for vertical resistors.

The reason for the high partial discharge values with horizontally installed resistors is that the field strengths act tangentially on the resistor tracks.

To produce the coaxial voltage divider shown in Fig. 4.81, the resistors were installed vertically as shown in Fig. 4.83. The position of the resistors influences the compensation. To keep this constant in series production, the position of the resistors was fixed by a meandering insulating part (see Fig. 4.84).

The voltage measurement systems, based on resistive-divider technology, are much more cost-effective and lighter than conventional inductive voltage transformers. In addition, they require only about 1/10 of the space compared to inductive voltage transformers. Figure 4.85 shows a coaxial divider installed in a medium voltage system. The trend in medium voltage switchgear is towards more and more integration and reduction of space, where the compensated R-dividers can contribute their part.

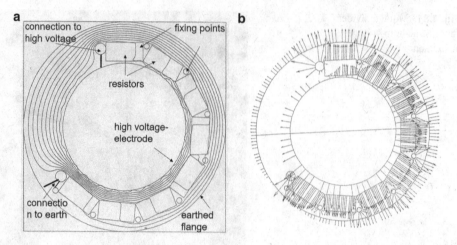

Fig. 4.82 Potential lines (**a**) and electric field strength (**b**) of the coaxial flange divider. The flat resistors are mounted horizontally

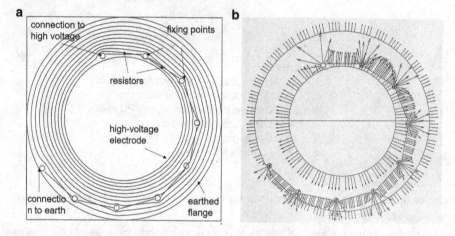

Fig. 4.83 Potential lines (**a**) and electric field strength (**b**) of the coaxial flange divider. The flat resistors are mounted vertically

4.8 RC Divider

4.8.1 Introduction

Over the last 100 years, inductive and capacitive voltage transformers have been well-proven for measuring voltage in transmission and distribution networks. The capacitive voltage transformer is used up to 90% for voltage measurement due to its reliability and low-cost production.

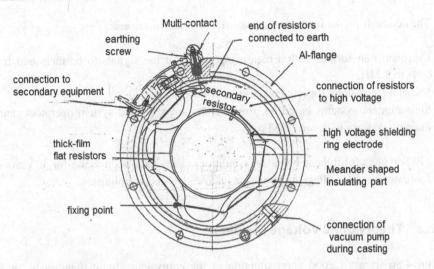

Fig. 4.84 Active part of the coaxial compensated R-divider for medium voltage gas insulated switchgear

Fig. 4.85 Coaxial compensated resistor divider in a gas insulated medium voltage switchgear

The current requirements of diagnostic, protection, and measurement technology for the transmission properties of capacitive and inductive voltage transformers cannot be met by these technologies. The RC voltage divider, which was developed in the 1960s, is suitable for these tasks.

The demands on the instrument transformers of the future are:

- Universal transformer with a measuring range for the signals to be measured from 0 Hz to 2 MHz.
- Use of a technology for low-cost voltage transformers.
- Measurement systems based on proven technology, as grid system operators cannot use new systems with increased risk.

The RC dividers fulfil these boundary conditions. They have been in use for 30 years for grid monitoring, DC voltage transmission and AC-DC-AC couplings.

4.8.2   Theory of RC Voltage Dividers

Figure 4.86 shows a good approximation of the equivalent circuit diagram of an RC divider, provided that the primary resistors R_{P_v} and the secondary resistors R_S are realised by a low-inductance design and the dynamic stress, caused by impulse voltages or insulator flashovers, does not change the ohmic resistance value.

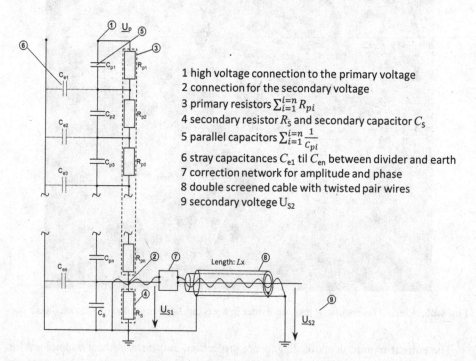

1 high voltage connection to the primary voltage
2 connection for the secondary voltage
3 primary resistors $\sum_{i=1}^{i=n} R_{pi}$
4 secondary resistor R_S and secondary capacitor C_S
5 parallel capacitors $\sum_{i=1}^{i=n} \frac{1}{C_{pi}}$
6 stray capacitances C_{e1} til C_{en} between divider and earth
7 correction network for amplitude and phase
8 double screened cable with twisted pair wires
9 secondary voltege U_{S2}

Fig. 4.86 Equivalent circuit diagram of an RC divider for high voltage [71]

The parallel capacitors C_{P_v} and the secondary capacitor C_S must also be of low inductance design, which requires modern connection technology for the capacitor elements connected in series.

For high reliability, the capacitors must have a partial discharge inception voltage (<5 pC) of 1.2 U_m, which is achieved by impregnating the capacitors under vacuum with environmentally friendly mineral oil or with SF_6 gas (gas pressure of several bar, hermetically sealed).

4.8.2.1 The Transmission Ratio of the RC Voltage Divider

From Fig. 4.86 follows the simplified equivalent circuit diagram in Fig. 4.87 of the RC divider without consideration of the stray capacitances.

The secondary voltage U_S as a function of the primary voltage U_P and the resistance and capacitance values R_P, R_S and C_P, C_S follow from the simplified equivalent circuit diagram in Fig. 4.87:

$$\frac{U_S}{U_P} = \frac{R_S}{R_S + R_P \frac{1+R_S \cdot j\omega C_S}{1+R_P \cdot j\omega C_P}} = \frac{C_P}{C_P + C_S \frac{1+1/j\omega C_S R_S}{1+1/j\omega C_P R_P}}$$

It follows from the equations that condition exists for the frequency independency of U_S/U_P is:

$$R_P \cdot C_P = R_S \cdot C_S$$

Furthermore, the following applies:

a) for $\omega \ll 1$: $\frac{U_S}{U_P} = \frac{R_S}{R_P + R_S}$,

b) for $\omega \gg 1$: $\frac{U_S}{U_P} = \frac{C_P}{C_P + C_S}$

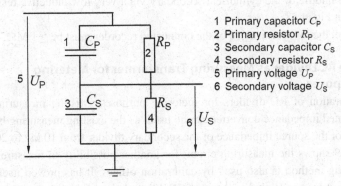

1 Primary capacitor C_P
2 Primary resistor R_P
3 Secondary capacitor C_S
4 Secondary resistor R_S
5 Primary voltage U_P
6 Secondary voltage U_S

Fig. 4.87 Simplified equivalent circuit diagram of the RC divider

4.8.2.2 The Influence of Temperature

The last two equations provide insight into the influence of temperature on the transmission ratio.

By introducing the linear dependence of resistances and capacitances on temperature applies:

$$\text{For low frequencies and } R_\text{P} >> R_\text{S}: \frac{U_\text{S}}{U_\text{P}} \approx \frac{R_{\text{S},20°\text{C}} \cdot (1 + \alpha_{\text{Rs}} \cdot \Delta\vartheta_{\text{Rs}})}{R_{\text{P},20°\text{C}} \cdot (1 + \alpha_{\text{Rp}} \cdot \Delta\vartheta_{\text{Rp}})}.$$

$$\text{For high frequencies and } C_\text{S} >> C_\text{P}: \frac{U_\text{S}}{U_\text{P}} \approx \frac{C_{\text{P},20°\text{C}} \cdot (1 + \alpha_{\text{Cp}} \cdot \Delta\vartheta_{\text{Cp}})}{C_{\text{S},20°\text{C}} \cdot (1 + \alpha_{\text{Cs}} \cdot \Delta\vartheta_{\text{Cs}})}.$$

To obtain a temperature-independent transmission ratio, the temperature coefficients of the resistors R_P and R_S and the temperature changes $\Delta\theta_{Rp}$ and $\Delta\theta_{Rs}$ of the corresponding elements must not show large differences.

The same requirements apply to the capacitors C_P and C_S. This is the prerequisite for the permissible changes of the divider ratio according to specified class accuracies.

Tests in a climatic chamber have confirmed the above conditions. For the design, this requirement means that the primary and secondary parts of the divider are exposed to the same temperatures.

4.8.2.3 The Connection System

The equivalent circuit diagram 4.86 shows the connection of the connecting cable (8) to the RC divider for quasi steady state processes at power frequency. This possibility corresponds to the traditional standard solution for inductive and capacitive voltage transformers. This connection is used for protection and metering purposes as well as for monitoring the power quality and for measurements of harmonics and subharmonics.

If the RC divider is used for diagnostic purposes, a 50 Ω coaxial cable must be used to connect the RC divider to the control room. This is ideally connected to the tap of the divider in the middle of the cylindrical secondary via a very low inductive resistor of 50 Ω as shown in Fig. 4.88.

In this case, the input resistance of the connected recorder must be ≥ 1 MΩ.

4.8.2.4 The RC Divider as Measuring Transformer for Metering Purposes

For the calibration of RC dividers for metering purposes, existing measuring bridges with a calibrated impedance converter can be used, as the existing measuring bridges are not suitable for the source impedance of the secondary divider from 10 kΩ to 20 kΩ.

Figure 4.89 shows the measuring circuit for another possibility to measure accuracy. This measuring method is also used by calibration offices. It has proven itself in practice with the use of two digital voltmeters (DVM) and a precision phase meter. The circuit can be operated both with an inductive voltage transformer calibrated by the PTB (Physikal.-Technische Bundesanstalt Braunschweig) and with a calibrated RC divider at a defined distance from the environment.

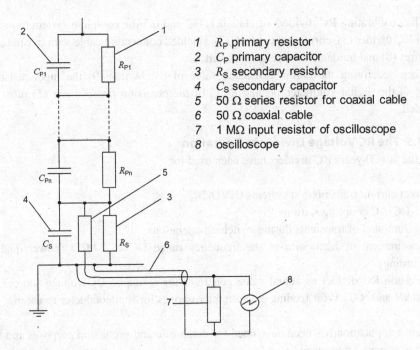

Fig. 4.88 Connection of an RC divider for high frequencies with a 50 Ω coaxial cable

1   R_P primary resistor
2   C_P primary capacitor
3   R_S secondary resistor
4   C_S secondary capacitor
5   50 Ω series resistor for coaxial cable
6   50 Ω coaxial cable
7   1 MΩ input resistor of oscilloscope
8   oscilloscope

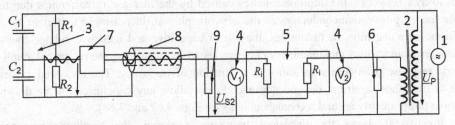

1   High voltage source
2   Calibrated voltage transformer
3   RC divider to be tested
4   Digital voltmeter (e.g., Agilent type 3440IA with input resistance R_{iVDM})
5   Precision phase meter (e.g., Krohn-Hite type 6620 with input resistance R_{iPhM})
6   Burden of the inductive voltage transformer
7   Correction network for amplitude and phase
8   Connection cable
9   burden (input resistance) of the secondary unit

Fig. 4.89 Calibration circuit for RC dividers for metering purposes

When calibrating RC dividers of class 0.1; 0.2 and 0.5 the complete system, consisting of RC divider (3), correction network (7), shielded connecting cable with twisted signal wires (8) and burden (9) must be measured.

When specifying the ohmic resistance value of the burden (9) the input resistors R_{iVDM} of the digital voltmeter (4) and R_{iPhM} of the precision phase meter (5) must be included.

4.8.2.5 The RC Voltage Divider in Operation
Over the last 20 years RC dividers have been used for

- Direct current transmission systems (HVDC)
- AC-DC-AC coupling stations
- Measurement of transients during switching operations
- Measurement of harmonics in the frequency range DC to 2 MHz (power quality recording)
- Precision RC divider as actual value generator for regulated DC voltage sources of 300 kV and 600 kV for feeding electron microscopes for semiconductor research.

The latest application has been developed for diagnostic and protection purposes and for metering at power frequency.

4.8.2.6 Use of the RC Dividers for Diagnostic Purposes
In today's networks, the additional losses caused by the increase in harmonics due to the use of power semiconductors in the network play an increasingly important role. To measure and monitor harmonics, the use of capacitive and inductive voltage transformers from various manufacturers [72] was tested by measuring the reciprocal amplitude response (transformation ratio) as a function of frequency. The measured values $\frac{U_P}{U_S} = f(\omega)$ show a strong dependency and do not allow any conclusions to be drawn about power quality around resonance points (see Figs. 4.67 and 4.68).

Figure 4.90 shows the measured frequency response, the amplitude response $\frac{U_S \cdot K_n}{U_P} = f(\omega)$ and the phase response $\Delta\varphi = \varphi_P - \varphi_S$ of an RC divider as a function of the frequency.

The measured frequency response in Fig. 4.90 shows the suitability of the RC divider for diagnostic purposes. The frequency response is stable up to the MHz range.

4.8.2.7 Determination of the Frequency Response
In the following some hints for measuring the frequency response of a 420 kV RC divider are given in the form of equations of the RC divider for $R_P \cdot C_P \neq R_S \cdot C_S$.

The transmission ratio is given by:

$$\frac{U_P}{U_S} = 1 + \frac{R_P}{R_S} \cdot \sqrt{\frac{1 + (R_S \omega C_S)^2}{1 + (R_P \omega C_P)^2}} \cdot e^{j(\varphi_P - \varphi_S)}$$

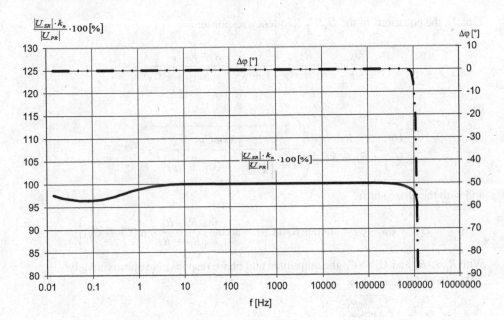

rated transmission ratio $K_r = U_{PR}/U_{SR} = 3600$
phase shift $\Delta\varphi = \varphi_P - \varphi_S$
rated primary voltage U_{Pr}
rated secondary voltage U_{Sr}

Fig. 4.90 Frequency response of a 420 kV RC divider

$\varphi_P - \varphi_S$ is the phase shift between the primary voltage U_P and the secondary voltage U_S.
The absolute value of the transmission ratio is:

$$\frac{U_P}{U_S} = 1 + \frac{R_P}{R_S} \cdot \sqrt{\frac{1 + (R_S\omega C_S)^2}{1 + (R_P\omega C_P)^2}}$$

with the phase shift

$$\varphi_P - \varphi_S = \arctan(R_S\omega C_S) - \arctan(R_P\omega C_P)$$

Or:

$$\frac{U_P}{U_S} = 1 + \frac{C_S}{C_P} \cdot \sqrt{\frac{1 + 1/(R_S\omega C_S)^2}{1 + 1/(R_P\omega C_P)^2}}$$

with the phase shift:

$$\varphi_P - \varphi_S = -\arctan\left(\frac{1}{R_S\omega C_S}\right) + \arctan\left(\frac{1}{R_P\omega C_P}\right).$$

Finally, the equations of the U_S/U_p frequency response:

$$\frac{U_S}{U_P} = \frac{R_S}{R_P + R_S} \cdot \frac{1 + R_P \cdot j\omega C_P}{1 + \frac{R_P \cdot R_S}{R_P + R_S} \cdot j\omega(C_P + C_S)}$$

or

$$\frac{U_S}{U_P} = \frac{R_S}{R_P + R_S} \cdot \sqrt{\frac{1 + (R_P\omega C_P)^2}{1 + \left[\frac{R_P \cdot R_S}{R_P + R_S} \cdot \omega(C_P + C_S)\right]^2}} \cdot e^{j(\varphi_P - \varphi_S)}$$

and with the phase shift:

$$\Delta\varphi = \varphi_P - \varphi_S = \arctan(R_P\omega C_P) - \arctan\left[\frac{R_P \cdot R_S}{R_P + R_S} \cdot \omega(C_P + C_S)\right]$$

With $R_P \gg R_S$ and $C_S \gg C_P$, the amplitude and phase response are approximately:
For large frequencies ω:

$$\frac{|U_S|}{|U_P|} \approx \frac{R_S}{R_P + R_S} \cdot \sqrt{\frac{1 + (R_P\omega C_P)^2}{1 + (R_S\omega C_S)^2}}$$

with the phase response:
$\Delta\varphi = \varphi_P - \varphi_S = \arctan(R_P\omega C_P) - \arctan(R_S\omega C_S)$.
or for small frequencies ω:

$$\frac{|U_S|}{|U_P|} \approx \frac{C_P}{C_S + C_P} \cdot \sqrt{\frac{1 + 1/(R_P\omega C_P)^2}{1 + 1/(R_S\omega C_S)^2}}$$

with the phase response:$\Delta\varphi = \varphi_P - \varphi_S = -\arctan\left(\frac{1}{R_P\omega C_P}\right) + \arctan\left(\frac{1}{R_S\omega C_S}\right)$.

The given equations are helpful for assessing the correction of the measured frequency response. Depending on the design of the RC divider, i.e., the capacitance and resistance values of C_P, C_S and R_P, R_S, and the range of the angular frequency ω, the given approximate formulae must be selected.

4.8.2.8 Measurement of the Amplitude and Phase Response of a 420 kV RC Divider

The illustration of the measurement of the frequency response in Fig. 4.91 shows the essential parameters of the measurement setup to compare measured amplitudes and phase responses. Figure 4.90 shows the frequency response of a 420 kV RC divider measured with it.

For reproducible frequency response measurements, a fixed measurement setup is a prerequisite. The area (6) must be as small as possible, as it is equivalent to an inductance in the measuring circuit and can strongly influence the result. The distance of the frequency generator to the RC divider was fixed at 2 m.

Fig. 4.91 Schematic representation of frequency response measurement

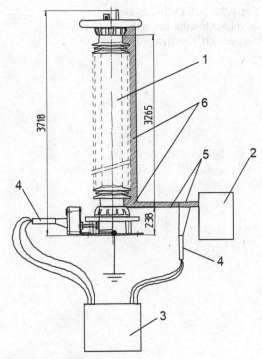

1   RC voltage divider 420 kV
2   Frequency generator
3   Oscilloscope, Tektronix Type TDS 30548
4   HF probes 10:1
5   Cu conductor 0.5 mm x 20 mm
6   This area must be kept as small as possible

4.8.3   Application of the RC Divider as Voltage Transformer for Protection Purposes and Energy Metering

Due to the worldwide dominant position of the capacitive voltage transformer (CVT) in outdoor switchgear for measuring high voltage, the behaviour of the CVT when used for distance protection devices [74–76] in the event of power disturbances was investigated. The complex transmission behaviour for transient processes requires additional engineering. The occurrence of short-term non-linear oscillations (ferroresonance) can lead to malfunctions of protective devices.

None of the papers mentions the problems of the capacitive voltage transformer and the C-divider, caused by trapped charges during short interruptions and reconnections (see Fig. 4.30). IEC standard 60044-7 of 1999 refers for the first time to the relationship between transient behaviour and trapped charges (adopted in IEC 61869-6:2016 [35]).

Fig. 4.92   420 kV RC divider
as voltage transformer in the
Singlewell/UK switchgear

With the system of the RC divider and protection device, the protection relay cannot malfunction due to the ideal transmission behaviour of the RC divider.

When using the RC dividers, no phenomena caused by trapped charges can occur.

The first 420 kV RC dividers were commissioned for the stations operated by the National Grid Company in the UK at Singlewell and Sellindge (see Fig. 4.92), both in the county of Kent, in the south-east of England (manufacturer Trench Switzerland Basel/CH).

These RC dividers are connected to multifunctional three-phase measuring instruments (diagnosis, protection, and measurement) of the ION 7500/ION 7600 type from Power Measurement, which comply with the standard DIN EN 50160 [77]. The main functions of the measuring devices are:

- Energy measurement
- Measurement of harmonics up to the 63rd harmonic (3150 Hz)
- Power Quality Measurement
- Measurement of voltage, current and frequency

The three-phase meters have an input resistance of 5 MΩ for the voltage signal and are directly connected to the secondary tap with the secondary voltage U_S.

The electrical data of the RC dividers used are:

• Highest voltage for the RC divider	$U_m = 420\ \text{kV}$
• Rated primary voltage	$U_{Pr} = 396/\sqrt{3}\ \text{kV}$
• Rated transmission ratio	$K_r = 3600$
• AC test voltage	$U_{eff} = 630\ \text{kV}$
• lightning impulse test voltage (1.2/50 μs)	$U_{peak} = 1425\ \text{kV}$
• Switching impulse test voltage (250/2500 μs)	$U_{peak} = 1050\ \text{kV}$
• Rated frequency	$f_r = 50\ \text{Hz}$
• Temperature class	$-50\ °\text{C} - +55\ °\text{C}$
• Rated secondary voltage	$U_{Sr} = 100/\sqrt{3}\ \text{V}$
• Partial discharge level to IEC 61869-1	<5 pC at 1.2 $U_m/\sqrt{3}$ <10 pC at 1.2 U_m
• Mechanical forces (horizontal and vertical)	3150 kN

4.8.4 Dimensioning of the RC Dividers

4.8.4.1 The Design for Outdoor Air Insulated Switchgear

RC dividers, as shown in Fig. 4.92, are produced in the entire high voltage range from 72.5 to 550 kV. Table 4.10 lists the main rated values of the main types of design.

The standard rated primary voltage U_{Pr} is $\frac{U_m}{1.2 \cdot \sqrt{3}}$. The resistance of the primary part R_P is determined so that a current of 300 μA flows at the rated primary voltage.

This value has proven itself in practice and is based on the heating in continuous operation at nominal voltage.

The capacitance of the parallel-connected capacitors C_P is determined by means of the stray capacitances against earth.

Measurements of influences have shown that the longitudinal capacitance C_P of the RC divider must be greater than 10 times the stray capacitance C_e so that the stray capacitance does not influence the measurement of the RC divider.

The stray capacitance of the RC divider is calculated as [78].

$$C_e = \frac{2 \cdot \pi \cdot \varepsilon_r \varepsilon_0 \cdot l}{ln\left[\frac{2l}{d} \cdot \sqrt{\frac{4h+l}{4h+3l}}\right]}$$

where:

ε_r relative dielectric constant of air $\varepsilon_r = 1$
ε_0 dielectric constant $\varepsilon_0 = 8854 \cdot 10^{-12}$ F/m
l length of the active part in m
d diameter of the active part in m
h height of the support structure in m

Table 4.10 Type series of RC dividers for the application as voltage transformers

Type	RCVT 71.5	RCVT 145	RCVT 245	RCVT 300	RCVT 420	RCVT 550
Maximum voltage U_m in kV$_{rms}$	72.5	145	245	300	420	550
AC test voltage U in kV$_{rms}$	140	275	460	460	630	680
Lightning impulse test voltage U_{BIL} in kV$_{peak}$	325	650	1050	1150	1425	1550
Switching impulse test voltage U_{SIL} in kV$_{peak}$				850	1050	1175
Smallest arcing distance in mm	630	1200	2000	2000	3200	3800
Standard creepage distance in mm	1813	3625	6125	7500	10,600	13,750
Rated primary voltage U_{Pr} in kV$_{rms}$	34.88	69.76	117.88	144.34	202.07	252.59
Resistance of the primary part R_p in MΩ	116	232	392	481	674	842

Using approximate equations, a simple formula for the calculation of C_e can be derived:

$$C_e = \frac{2 \cdot \pi \cdot \varepsilon_r \varepsilon_0 \cdot l}{ln\left[\sqrt{2} \cdot \frac{l}{d}\right]}$$

Figure 4.93 shows the dimensions on which the calculation is based.

The calculation of the stray capacitance of a 420 kV RC divider according to the two formulas given above:

Fig. 4.93 Dimensions of a 420 kV RC divider with support structure (see Fig. 4.92)

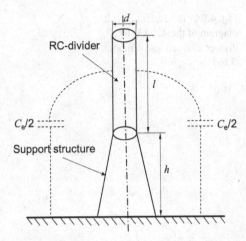

Numerical values:
$l = 3.2$ m
$h = 3.0$ m
$d = 0.25$ m

a) $C_e = \dfrac{2 \cdot \pi \, 28.854 \cdot 10^{-12} \cdot 3.2}{ln\left[\frac{2 \cdot 3.2}{0.25} \cdot \sqrt{\frac{4 \cdot 3 + 3.2}{4 \cdot 3 + 3 \cdot 3.2}}\right]} F = \dfrac{178.02 \cdot 10^{-12}}{ln \, 21.45} F = 58.04$ pF

b) $C_e = \dfrac{2 \cdot \pi \cdot 8.854 \cdot 10^{-12} \cdot 3.2}{ln\left[\sqrt{2} \cdot \frac{3.2}{0.25}\right]} F = \dfrac{178.02 \cdot 10^{-12}}{ln\left[\sqrt{2} \cdot \frac{3.2}{0.25}\right]} F = 61.47$ pF

The approximate formula in b) gives a slightly higher value. For the further considerations, the value of the total stray capacitance of the RC divider against earth of $C_e = 60$ pF is used.

Figure 4.94 shows the equivalent circuit diagram of a 420 kV RC divider with the longitudinal capacitances C_{Pi} and the stray capacitances C_{ei} to earth.

The active parts of the RC dividers are manufactured in modules. The divider for 420 kV in Fig. 4.94 is divided into 4 modules.

The numerical values of the 420 kV RC divider refer to the rated voltage:
$U_{Pr} = \dfrac{U_m}{12 \cdot \sqrt{3}} = \dfrac{420 \text{ kV}}{1.2 \cdot \sqrt{3}} = 202.07$ kV.

Primary resistance $R_p = 673.58$ MΩ, equivalent to 168.4 MΩ per module Primary capacitance $C_p = 810$ pF, equivalent to 3240 pF per module Stray capacitance $C_e = 60$ pF, equivalent to 15 pF per module

The chain network can be optimised with the LaPlace transformation according to Doetsch [79].

As mentioned above, the longitudinal capacitance of the primary capacitors C_p should be at least 10 times the stray capacitance C_e. In the above example of the 420 kV RC divider, the longitudinal capacitance is a factor of 14 greater than the stray capacitance, resulting in a better frequency response.

Fig. 4.94 Equivalent circuit
diagram of the 420 kV RC
divider, divided into 4 modules
(I to IV)

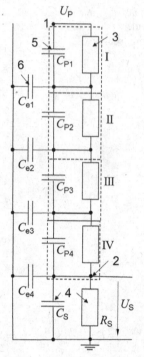

1   High voltage connection to the primary voltage U_P
2   Connection for secondary voltage U_S
3   Primary resistors R_P
4   Secondary resistor R_S and secondary capacitor c_S
5   Primary capacitors C_{Pi}
6   Stray capacitances against ground C_{ei}

In another example of a 245 kV RC divider the calculated stray capacitance is 40 pF. For the capacitance of the primary capacitor C_p a capacitance of 1492 pF was chosen, which makes it a factor of 40 greater than the stray capacitance. The influence of the stray capacitance on the RC divider is therefore very small. The frequency response of this 245 kV RC divider is significantly extended compared to the 420 kV RC divider.

Figure 4.95 shows an active part of an RC divider with the stack of capacitor elements for C_p and the resistors connected in parallel. For the capacitor elements, a mixed dielectric of paper and polypropylene films is used for oil-insulated dividers, while pure polypropylene film is used as dielectric for SF_6 insulated dividers.

The design development of the high-voltage RC dividers is closely linked to the technological development of the resistors and capacitors.

Fig. 4.95 Active part of an
RC divider

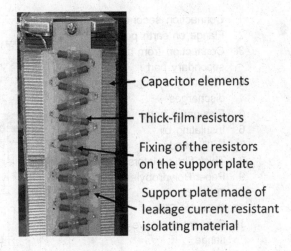

Capacitor elements

Thick-film resistors

Fixing of the resistors
on the support plate

Support plate made of
leakage current resistant
isolating material

4.8.4.2 RC Dividers in Gas-Insulated Switchgear
4.8.4.2.1 The Three-Phase Encapsulated RC Divider with Oil-Impregnated Dielectric

Figure 4.96 shows a section of an RC divider with oil-impregnated mixed dielectric of paper-polypropylene. Three such dividers are assembled to a three-phase encapsulated RC divider for GIS. The data for these dividers are:

• Maximum permissible voltage	$U_m = 145$ kV
• AC test voltage	$U_{rms} = 275$ kV
• Lightning impulse test voltage (BIL)	$U_{peak} = 650$ kV
• Arcing distance:	410 mm
• Diameter of the RC divider (single-phase):	108 mm
• Wall thickness FRP tube	4 mm
• Weight	6 kg

For the stability of the primary capacitance C_P, the stack of capacitor elements is pressed with springs (pos. 16 and 17). Some length measurements were entered in Fig. 4.96 to give an idea of the size.

The arcing distance is 490 mm -(2 · 40 mm) = 410 mm, which corresponds to a longitudinal field strength of 275 kV/410 mm = 0.67 kV/mm at a test voltage of 275 kV.

To avoid partial discharges on the insulating surface position 4 in Fig. 4.96 was designed as shown in Fig. 4.97.

Figure 4.98 shows the structure of the three active parts for installation in the GIS pressure housing. Figure 4.99 shows the three active parts installed in the GIS enclosure.

1. Connection secondary part
2. Flange on earth potential
3. Connection from primary to secondary part
4. Undercut to avoid sliding discharges
5. Expansion body
6. Insulating oil
7. FRP cylinder
8. Thick film resistors
9. Paper/Polypropylene Dielectric
10. Potential Rings
11. Connection active part - flange
12. Double seal
13. Flange on high voltage
14. Bushing
15. Gas filling device
16. Screws for pressing the capacitor windings
17. Springs for pressing the capacitor windings

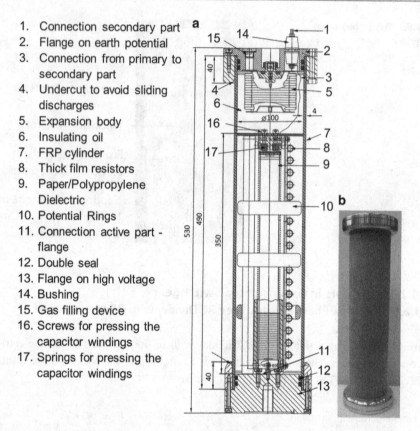

Fig. 4.96 Active parts of a 145 kV RC divider with oil-impregnated dielectric for a three-phase encapsulated GIS without secondary part **a** Sectional view **b** assembled divider

Fig. 4.97 Detail of flange design (position 4 in Fig. 4.96) to prevent sliding discharges on the insulation surface

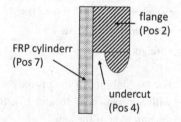

The resistor part in the primary part of the RC divider is realised by thick-film resistors mounted on a FRP support plate, as shown in Fig. 4.100. This resistor plate is mounted parallel to the capacitor windings in the FRP tube (see Fig. 4.101).

The resistors are dimensioned so that the current through the resistors is about 0.3 mA.

For field control, potential rings (item 4 in Fig. 4.101) are installed in the active parts, which are connected to the capacitor column and the resistors. The potential rings are

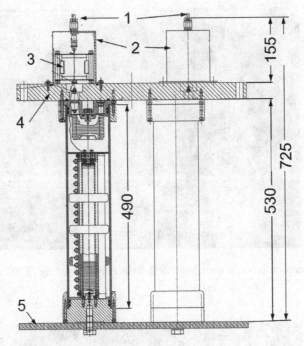

1. Connection for output voltage U_s
2. Housing of the secondary part
3. Secondary resistors and capacitors arranged coaxially
4. Flange of GIS pressure vessel on earth potential
5. Mounting plate RC divider for transportation

Fig. 4.98 Active parts of the three-phase RC divider with the secondary parts of the divider mounted on the cover of the GIS pressure vessel

coated with insulating varnish to increase the breakdown voltage between the rings. The capacitor column is held by FRP guide rails and pressed by springs.

The value of the primary capacitance C_P of the RC divider should be at least 10 times greater than the mutual stray capacitances C_{st}, as well as the stray capacitances relative to the earthed vessel C_{sE} (see Fig. 4.102).

Figure 4.103 shows schematically the complete 3-phase RC divider, with the high voltage connections of phases R, S and T on the GIS bushing, the pressure housing, and the secondary parts. (Note: the design can also be used without GIS bushing).

Figure 4.104 shows a photo of the three active parts ready for installation in the pressurised enclosure. It shows especially the connection to the GIS bushing.

The secondary part of the RC-divider is coaxial and is located outside the pressure vessel.

Figure 4.105 shows the structural design of the coaxial secondary part, which is designed as shown in the schematic diagram in Fig. 4.88 in Sect. 4.8.2.3. The connecting

Fig. 4.99 Connection point of three-phase RC divider to the bushing (3 RC dividers in an SF$_6$ pressure tank)

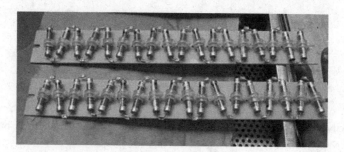

Fig. 4.100 Primary resistor part of the RC divider

plate (pos 1) is supported by two insulating supports (pos 2). In the middle the connection with 50 Ω (pos 3) between the tapping of the divider and the secondary connection is shown.

The secondary part in Fig. 4.105 consists of 8 capacitors (pos 4), two thick-film resistors (pos 5) and a gas protective spark gap (pos 6, not visible). The upper plate (pos 7) is connected to earth with the housing of the secondary part (Fig. 4.106).

The housings of the secondary parts are screwed on the outside of the cover of the pressure vessel for the primary. Figure 4.107 shows the complete RC-divider with mounted secondary parts.

For the described three-phase RC divider for $U_m = 145$ kV the field strengths were calculated for an increased test voltage of 330 kV. (1 min test voltage according to IEC is 275 kV). Figure 4.108 shows a schematic representation of the calculated arrangement.

Fig. 4.101 Active parts of the RC dividers before installation in the FRP cylinders

1. Pressing device for the capacitor stack
2. Primary resistors
3. Capacitor stacks (flat windings connected in series)
4. Potential Rings
5. FRP guide rail

Fig. 4.102 Equivalent circuit diagram of the three-phase RC divider for the primary capacitance C_P and the stray capacitances C_{st} and C_{sE}

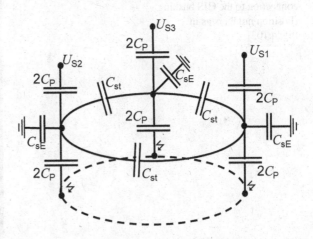

Figure 4.109 shows the calculated electric fields between the three RC dividers at a voltage of 330 kV rms. The applied voltages were entered in the calculation programme according to the phase position of the three RC dividers for a three-phase system. The current phase positions are shown in the figures. The dimensions of the flange at high voltage potential are shown in Fig. 4.109d. The dimensions of the electrodes are shown in Fig. 4.109f. The geometrical arrangement of the electrodes is shown in Fig. 4.109b.

Figure 4.109a. shows the field strength between the lower flanges. The maximum field strength is 15 kV/mm. Figure 4.109c. shows the field strength between the first

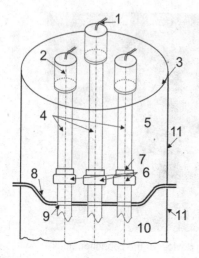

1. Connection for output voltage of the secondary part
2. Coaxial secondary part
3. Cover of the primary gas compartment
4. Primary parts
5. Gas space of the primary part
6. Connectors between primary parts and GIS bushing
7. Flange of the primary part, screwed to pos. 6
8. GIS bushing
9. Cast-in screw inserts (bushings)
10. Gas room of the GIS plant
11. Pressure vessel

Fig. 4.103   Three phase RC divider

Fig. 4.104   Three-phase connection to the GIS bushing (Position numbers as in Fig. 4.103)

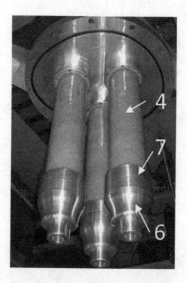

intermediate electrodes at a voltage of ¾ · 330 kV = 247.5 kV. The maximum field strength here is 7.3 kV. Figure 4.109e. shows the maximum field strength against the earthed wall of the pressure housing, it is 14 kV/mm. The phase positions of the RC dividers at which the maximum field strengths are obtained are shown in the figures.

Figure 4.110 shows the calculations on the radii of the flanges and electrodes. Figure 4.110a shows the field image of the whole arrangement, Fig. 4.110b and c show the details at the radius of the first intermediate electrode, and Fig. 4.110d and e show the details at the radius of the high voltage flange of the RC divider.

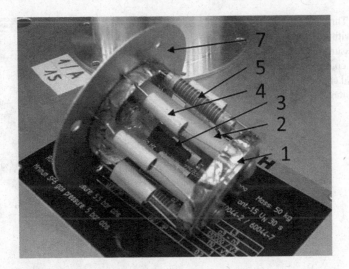

Fig. 4.105 Active part of the secondary part

Fig. 4.106 Coaxial secondary of RC divider with housing

In summary, it must be said for the described three-phase RC divider that it is a proto-type with oil-impregnated mixed dielectric, which has not yet been optimised for a cost-effective solution.

The present design fulfils all conditions of a three-phase inductive voltage trans-former, such as class accuracy 0.2% and the protective functions. However, the system is much smaller in volume and less expensive, it has excellent dynamic properties, and it can be used for diagnostic purposes of the gas-insulated switchgear, which is not pos-sible with an inductive or capacitive voltage transformer.

Fig. 4.107 Three-phase
RC divider with secondary
(without pressure vessel).
1 housing of coaxial secondary
part, 2 coaxial connector for
secondary voltage

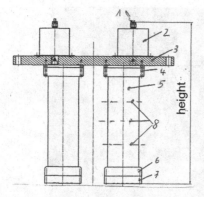

Pos.1   Connection secondary voltage

Pos.2   Housing Secondary part

Pos.3   Cover of pressure vessel at earth
        potential

Pos.4   Flange primary part at earth
        potential

Pos.5   Primary part

Pos.6   Flange on high voltage

Pos.7   Connection piece

Pos.8   Potential rings in the FRP housing
        of the divider

Fig. 4.108 Schematic representation of the three-phase RC divider for field strength calculation

4.8.4.2.2 The Single-Phase Encapsulated RC Divider with SF$_6$ Gas Impregnated Dielectric

Figure 4.111 shows a 145 kV RC divider (pos 1) with integrated secondary part (pos 2). The voltage between the potential rings (pos 3) is $1/\sqrt{3} \cdot 145/5$ kV $= 16.7$ kV. In this divider, polypropylene (PP) is used as the dielectric of the capacitors. Similar results are achieved using polyethylene (PE). RC dividers with SF$_6$ impregnated dielectric must be installed in a separate gas chamber. After impregnation, the gas compartment must not be opened.

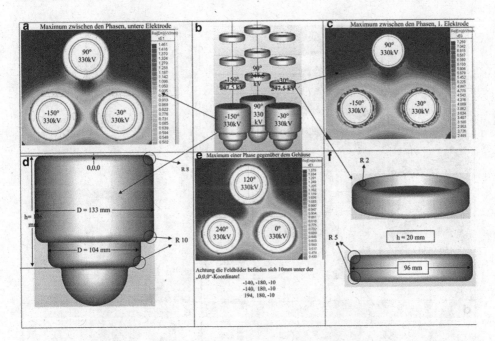

Fig. 4.109 Calculations of the field strength between the three RC dividers, and against the pressure housing

To facilitate gas impregnation between the films, the guide rails were replaced by narrow strips, as shown in Fig. 4.112. The partial discharge inception voltage is only determined by the gas gap between the foils and the gas pressure.

The practical impregnation of the capacitors with SF_6 takes place in the following steps:

1. Set the flat winding capacitors under high vacuum (<0.01 mbar for 2 h)
2. Then fill with SF_6, pressure 6 bar, duration 12 h
3. Then reduce to operating pressure 5.5 bar absolute.

The structure of the secondary part is shown in Fig. 4.113. It consists of flat windings and an additional round winding balancing capacitor with the dielectric polystyrene.

Figure 4.114 shows the influence of temperature on the divider ratio of the RC divider. The divider meets the requirements of the accuracy class 0.5 up to 40 °C. Several temperature cycles were run to obtain the curves.

4.8.4.2.3 The Three-Phase Encapsulated RC Divider with SF_6 Impregnated Dielectric

The structure of the three-phase encapsulated RC divider with SF_6 impregnation is similar to the single-phase encapsulated RC divider described above. Polypropylene film

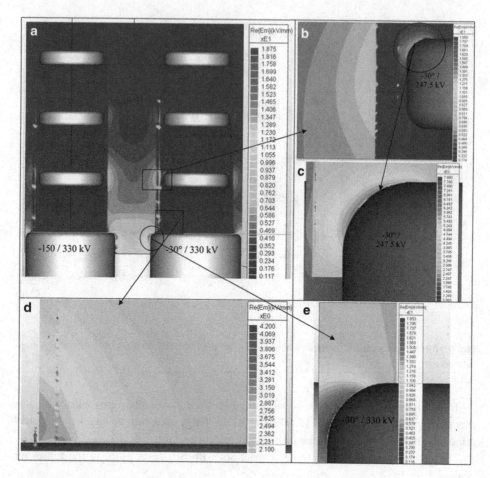

Fig. 4.110 Calculation of the field strengths at the radii of the electrodes

was used as dielectric. The divider for $U_m = 145$ kV in Fig. 4.115 is divided with 7 electrodes. The voltage between the electrodes is $1/\sqrt{3} \cdot 145/8 = 10.5$ kV.

Figure 4.116 shows the three-phase RC divider in the pressure vessel of the GIS. The secondary part of the divider is mounted outside the pressure vessel in a separate housing. Figure 4.117 shows a sectional view of the divider.

4.8.4.2.4 The RC Divider with Ceramic Capacitors and Integrated Thick-Film Resistors

European Patent EP098003B1 "RC voltage divider" [80] describes another variant of RC divider for use in GIS. This variant consists of ceramic capacitors on whose outer surface the resistors are printed in thick-film technology. Several such RC modules, as shown in Fig. 4.118, are stacked on top of each other, each separated by electrodes for external field control.

Fig. 4.111 The single-phase RC divider with SF_6 impregnated dielectric

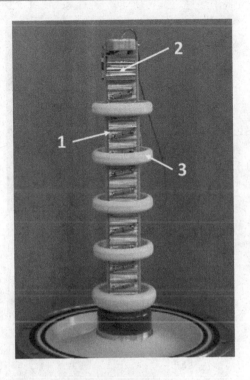

Fig. 4.112 Partial view of the RC divider with the narrow guide rails

1: Flat film wrapping with
   SF$_6$ impregnation

2: Springs for pressure regulation
   of the capacitor stack

3: Balancing capacitor

Fig. 4.113 Secondary part of the RC divider

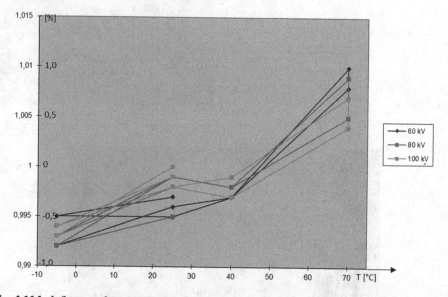

Fig. 4.114 Influence of temperature on the divider ratio and RC divider error. the right scale on the ordinate shows the percentage error, the left scale shows the relative transmission ratio

For version a., where the resistors are applied in meandering form, attention must be paid to the voltage load between the meanders, as sliding discharges on the ceramic surface can lead to flashovers.

Special glass ceramic plates were used for the ceramic capacitors. The following parameters apply to these ceramics, which are acceptable for use as capacitors:

- Dielectric strength > 40 kV/mm
- Dielectric constant $\varepsilon_r = 200 \dots 300$
- Dielectric losstan $\delta = 1{,}5 \cdot 10^{-2}$ (at T = 20 °C)

Fig. 4.115 Active parts of the three-phase encapsulated RC divider with SF$_6$ dielectric for 145 kV

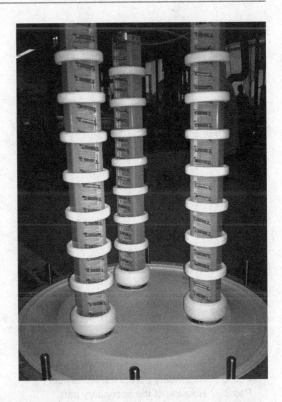

Fig. 4.116 3-phase encapsulated RC divider for 145 kV **a.** The active parts of the RC divider in the pressure housing of the GIS **b.** The assembled three-phase encapsulated RC divider

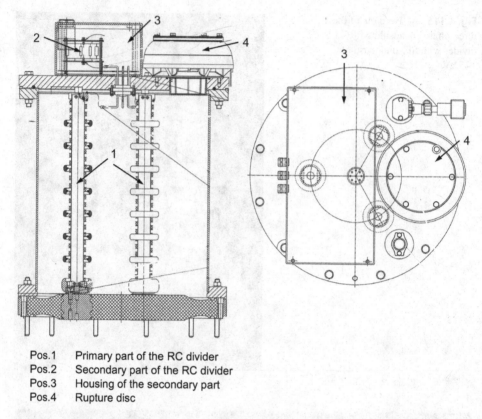

Pos.1   Primary part of the RC divider
Pos.2   Secondary part of the RC divider
Pos.3   Housing of the secondary part
Pos.4   Rupture disc

Fig. 4.117   Sectional view of a three-phase encapsulated RC divider for 145 kV GIS

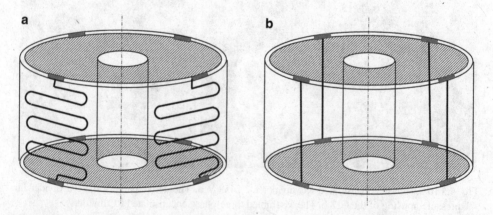

Fig. 4.118   RC module consisting of a ceramic capacitor with printed resistors. **a.** Resistors in meandering form; **b.** Resistors as straight lines

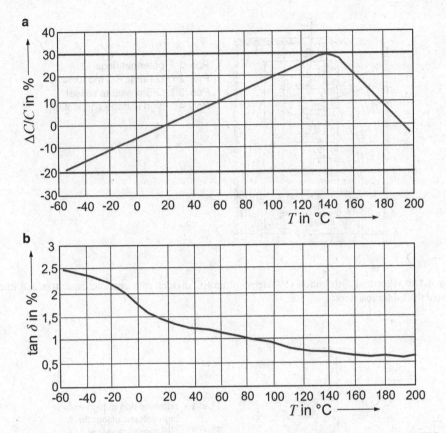

Fig. 4.119 Glass ceramic properties as a function of temperature **a**. Capacity change $\Delta C/C$; **b**. Dielectric losses tan δ

- Capacity change as a function of temperature according to Fig. 4.119a
- Change of dielectric losses with temperature according to Fig. 4.119b

The thick-film resistors can be printed and fired onto the cylindrical surface as vertical lines or in meandering form (see Fig. 4.118).

Figure 4.120 shows such an RC divider, installed in a GIS housing, with the potential lines in 5% steps. This divider consists of 6 ceramic RC modules and has 5 intermediate electrodes. It is designed for $U_m = 145$ kV with a test AC voltage of 275 kV.

Figure 4.121 shows the field calculation with the field strengths at the electrodes.

At point ①, the field strength between the potential ring and the pressure housing can be calculated according to the cylinder formula $\left|\vec{E}\right| = \frac{U}{r \cdot \ln(r_a/r_i)}$. For the test voltage of 275 kV, this results in a field strength for the RC divider in Fig. 4.121 of

$$\left|\vec{E}\right| = \frac{175 \text{ kV}}{42 \text{ mm} \cdot \ln(115 \text{ mm}/42 \text{ mm})} = 6.5 \text{ kV/mm}.$$

The vector has the highest field strength at the point ② with 11 kV/mm.

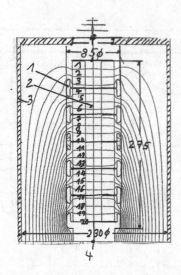

Fig. 4.120 Potential distribution (5% steps) of an RC divider with ceramic capacitors and integrated thick-film resistors

Pos. 1   potential rings
Pos. 2   ceramic RC modules
Pos. 3   GIS pressure vessel
Pos. 4   high voltage connection

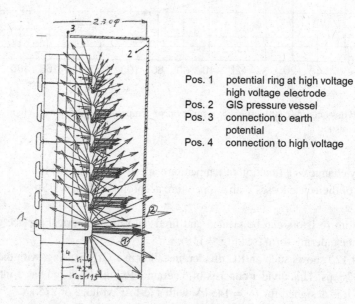

Pos. 1   potential ring at high voltage
         high voltage electrode
Pos. 2   GIS pressure vessel
Pos. 3   connection to earth
         potential
Pos. 4   connection to high voltage

Fig. 4.121 Field strength calculation on the RC divider electrodes

4.9 Optical Voltage Transformer (Pockels Effect)

4.9.1 Principle of Optical Voltage Measurement with the Pockels Effect

Like the Faraday effect for optical current measurement, the Pockels effect can be used for optical voltage measurement. The Pockels effect, also known as linear electro-optical effect, changes the birefringence in a crystal when an electric field is applied. This change in birefringence alters the polarisation state of the light as it passes through the crystal. This change in polarisation can then be analysed and gives a measure of the voltage across the crystal. Figure 4.122 shows this principle of the Pockels effect.

In contrast to the Faraday effect for current measurement, however, the Pockels effect does not occur in glass fibres, but only in special glass crystals. This restricts the use of the Pockels effect for direct voltage measurement. Two possible solutions are described below.

4.9.2 The Design of Optical Voltage Transformers

To use the Pockels effect for the measurement of high voltages, two different versions exist:

- The electric field is measured at one or more points and the voltage is calculated.
- A large crystal is used to bridge the entire voltage.

Figure 4.123 shows the realisation of a voltage transformer using several crystals which, within an insulator, measure the electric field at different points. In the electronic unit of these optical voltage transformers, the total voltage is calculated by combining these field measurements [81].

With this design, care must be taken to ensure that external electric fields from adjacent phases, or field distortions due to earthed constructions in the vicinity of the optical voltage transformer do not influence the result of the voltage measurement.

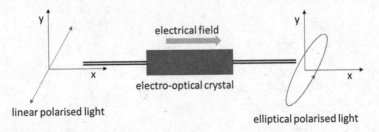

Fig. 4.122 Principle of the Pockels effect for optical voltage measurement

Fig. 4.123 Optical voltage transformer with three electric field sensors according to the Pockels effect [81]

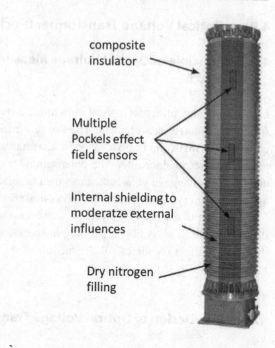

composite insulator

Multiple Pockels effect field sensors

Internal shielding to moderatze external influences

Dry nitrogen filling

Figure 4.124 shows the realisation of an optical voltage transformer by using a large crystal [82] that bridges the entire voltage.

The production of such large crystals is not easy and therefore associated with high costs. To maintain the full voltage across the crystal, it must be placed between two electrodes, which are in a gas-insulated pressurised insulator. To prevent the electric field strength in air outside the insulator from becoming too high, the insulator must have a certain minimum diameter.

With the RC divider described in Sect. 4.8, a technology is available which does not have the disadvantages in the realisation of the optical voltage transformer and can be connected directly to the protection and measuring devices without additional complex electronics. The authors therefore prefer the RC divider over the optical voltage transformer.

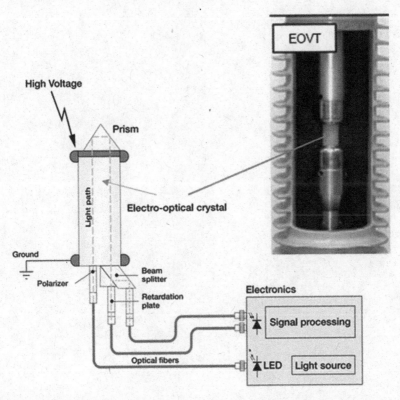

Fig. 4.124 Optical voltage transformer using a crystal to bridge the entire voltage [83]

Quality

5

5.1 Gas in Oil Analyses

5.1.1 Introduction

To assess the quality of the insulation of an instrument transformer during operation, the analysis of the dissolved gases in the insulating oil is a frequently applied technique. In the event of insulation faults which lead to partial discharges, or when exposed to high temperatures, the oil in the insulation decomposes and gases such as hydrogen and hydrocarbons are formed which are dissolved in the oil. If the transformer was not kept under sufficient vacuum during impregnation, the oil contains high levels of nitrogen and oxygen. Partial discharges, which affect the insulating paper, produce carbon monoxide, and carbon dioxide in addition to the hydrocarbons. More information can be found in the IEC standard IEC 60599 [84].

Dissolved gases do not immediately lead to failure of the transformer but can be an indication of defects or overheating. It is therefore recommended to take an oil sample every 10 years and to check for dissolved gases.

If gas is continuously formed due to a defect in the instrument transformer, this can lead to a pressure increase in the transformer and thus to an expansion of the oil expansion bellows. The final consequence is a breakdown in the instrument transformer. A certain increase in dissolved gases is not harmful to the instrument transformer and usually occurs over time, but if the gas concentration rises above the limit of the normal values

© Springer Fachmedien Wiesbaden GmbH, part of Springer Nature 2022
R. Minkner and J. Schmid, *The Technology of Instrument Transformers*,
https://doi.org/10.1007/978-3-658-34863-2_5

Table. 5.1   Limit values for gases dissolved in the oil of an instrument transformer

Gas type	concentrations in new transformers vpm or µl/l	Normal concentration in vpm or µl/l	Limit values at which the transformer must be taken out of operation in vpm or µl/l
Hydrogen H_2	<30	<300	<1000
Carbon monoxide CO		<300	<1000
Carbon dioxide CO_2		<900	<2000
Methane CH_4	<5	<30	<75
Ethane C_2H_6	<5	<25	<70
Ethylene C_2H_4	<1	<4	<25
Acetylene C_2H_2	not detectable	<1	<5

(3rd column in Table. 5.1), it must be investigated further. It is then advisable to carry out new gas-in-oil analyses after a few months to detect a possible further increase [85]. The values in the 3rd column in Table. 5.1 are lower for ethane, ethylene, and acetylene than the recommended limit values in IEC 60599 Annex A3 [84].

If the concentration of a gas exceeds the limit values in column 4 of Table. 5.1, it must be taken out of operation [85].

In newly produced instrument transformer, the concentration of the gases should not exceed the values in the 2nd column of Table. 5.1.

5.1.2   Sampling

When taking oil samples for gas-in-oil analysis, care must be taken to ensure that the samples are not contaminated and that no gas can escape from the oil. For this purpose, a glass syringe is attached to the oil drain valve of the instrument transformer, which is first rinsed with insulating oil before the sample is taken. Under no circumstances may the sample come into contact with air. Otherwise, the possibly dissolved hydrogen will escape from the oil very easily, therefore the hydrogen content can be falsified if the sample is taken incorrectly. The sample should only be taken by trained specialists. Gas-tight glass syringes are suitable for taking the samples. However, these should first be rinsed out with the insulating oil.

As the instrument transformers have small oil volumes, it is absolutely necessary to refill the oil that has been removed. For this purpose, there are devices which allow the samples to be taken and the oil to be refilled without the sample coming into contact with

Fig. 5.1 Device for extracting and refilling oil from oil-paper insulated instrument transformer

air or air entering the instrument transformer. Figure 5.1 shows a suitable device for taking the oil sample and replenishing the extracted oil.

5.1.3 Interpretation of Gas-in-Oil Analyses

Depending on the type of fault in the instrument transformer, different gases occur. This allows to determine the type of fault in the instrument transformer. Partial discharges with a low energy level result in an increase of the hydrogen content in the insulating oil accompanied by methane. In the case of high-energy discharges, acetylene typically occurs in addition to hydrogen, accompanied by methane and ethylene. If the discharges are in the oil-paper insulation, carbon monoxide and carbon dioxide also occur during discharges in the instrument transformer.

Overheating of the oil is characterised by the presence of methane and ethane at temperatures of 150 to 300 °C. At even higher temperatures (500–800 °C) ethylene occurs together with hydrogen, methane, and ethane and at very high temperatures of over 1000 °C hydrogen, ethylene and acetylene occur together with methane.

Table. 5.2 lists the types of faults and the gases occurring

Table. 5.2   Gas evolution in oil-filled instrument transformers

	partial discharge with low energy		Powerful discharges		Overheating of the oil		
	Oil	oil/paper	Oil	oil/paper	150 – 300 °C	500 – 800 °C	> 1000 °C
Hydrogen H_2	●	●	●	●		○	●
Methane CH_4	○	○	○	○	●	○	○
Ethane C_2H_6					●	○	
Ethylene C_2H_4			○	○		●	●
Acetylene C_2H_2			●	●			●
Carbon monoxide CO		○		●			
Carbon dioxide CO_2		○		○			

● characteristic gases, ○ accompanying gases.

5.2   Assessment of PD Results

The partial discharge measurement, which is carried out as a routine test on each instrument transformer, shows the quality of the insulation of the transformers. According to the standard [33], limits of the measured partial discharge must be observed. These are for oil-paper insulation 5 pC at $1.2U_m/\sqrt{3}$ and 10 pC at $1.2\ U_m$ for insulated systems and at U_m for earthed systems.

If partial discharges occur, a more detailed analysis of the measurements allows an indication of the cause of the partial discharges. For this purpose, the phase position of the PD pulses, the frequency and regularity of the pulses, the change in intensity with the applied voltage and the ratio of the inception and extinction voltage are used as parameters for evaluation [10].

Modern analysis equipment for measuring partial discharges records the PD pulses, thus creating a picture of the intensity and frequency of the pulses over the phase position. [86], [87].

In the following, different partial discharge causes and their typical partial discharge parameters are shown [88], [10]:

Discharge at Metallic Electrode in Air (Corona Discharge)
Partial discharge pulses around one of the two peaks with equal intervals. If the electrode is at high voltage, the pulses occur at the negative half-wave (see Fig. 5.2 a.), if the electrode is at earth potential, the pulses occur at the positive half-wave. With increasing voltage, the number of pulses increases, but their amplitude remains the same. The PD inception voltage is equal to the PD extinction voltage (see Fig. 5.2 b.)

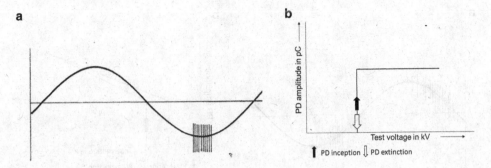

Fig. 5.2 Corona discharge in air at a metallic tip at high voltage potential **a**. Partial discharge image, **b**. Amplitude as a function of voltage

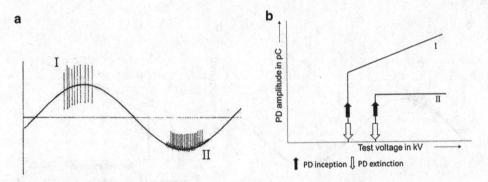

Fig. 5.3 Discharge at a metallic tip at high voltage potential under oil **a**. Partial discharge pattern, **b**. Amplitude as a function of voltage

Discharge at Metallic Electrode Under Oil

Partial discharge pulses at both peaks with equal distances. In one half-wave more pulses with the same amplitude, which remains the same when the voltage increases (II). At the other half-wave there are fewer pulses with higher amplitude, which increases as the voltage rises (I) (see Fig. 5.3).

Discharge at Defects in the Dielectric

Partial discharge pulses on both half-waves before the peak. The amplitudes can differ up to a factor of 3. If there are several faults, the amplitude of the pulses increases with increasing voltage [see Fig. 5.4]. When the voltage is reduced, the partial discharges stop at lower voltage than they started.

Discharge along Insulation Surfaces (Sliding Discharge)

Partial discharge pulses at both half-waves before the peak value, often with amplitude increasing towards the peak value. With increasing voltage, the amplitude of the pulses

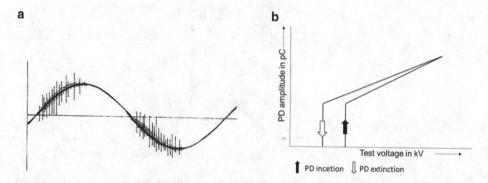

Fig. 5.4 Partial discharge at a defect in the dielectric, e.g. cavities or unimpregnated areas **a**. Partial discharge image, **b**. Amplitude as a function of voltage

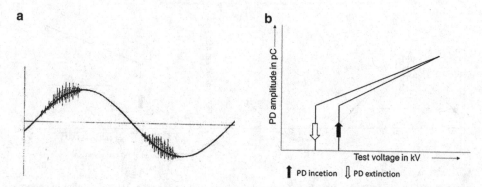

Fig. 5.5 Partial discharge along an insulating surface **a**. Partial discharge image, **b**. Amplitude as a function of voltage

increases (see Fig. 5.5). When the voltage is reduced the partial discharges stop at lower voltage than they started.

Discharge at Poor Electrical Contacts

Irregular partial discharge pulses symmetrical to the zero crossings. The amplitudes increase with the voltage (see Fig. 5.6). If the contact is "welded" at higher voltage the partial discharge can also suddenly disappear (see Fig. 5.6 b. case II).

Discharges on Metal Parts at Free Potential

Partial discharge pulses on both half waves, often occurring in pairs. The pulses can wander across the image. As the voltage rises, the number of pulses increases and the gaps between the pulses become smaller, but the amplitude of the pulses remains the same (see Fig. 5.7).

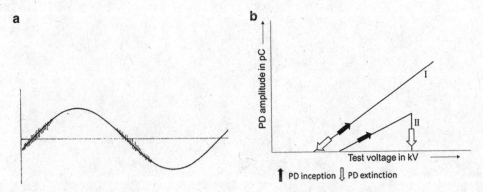

Fig. 5.6 Partial discharge on a poor electrical contact **a**. Partial discharge image, **b**. Amplitude as a function of voltage

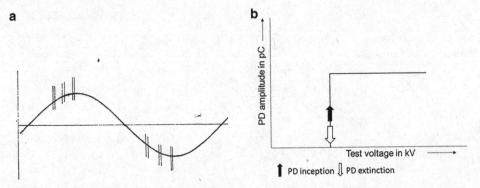

Fig. 5.7 Partial discharge on a metallic electrode at free potential **a**. Partial discharge image, **b**. Amplitude as a function of voltage

The determination of the cause of occurring partial discharges can often be difficult due to overlapping of several partial discharge sources.

Partial discharges in oil-paper insulated instrument transformers can be caused by insufficient impregnation of the paper and the resulting air pockets in the insulation. In such cases, post impregnation can help to eliminate the partial discharges. This can be done by warming up the insulation or by increasing the pressure in the insulation for a certain time.

Bibliography

1. Rudolf Bauer; Die Messwandler; Springer Verlag Berlin Heidelberg;1953
2. Ingmar Grambow; Messwandler für Mittel- und Hochspannungsnetze; Technische Akademie Esslingen Kontakt und Studium Band 554; Expert Verlag Renningen, 2000
3. Franz Ollendorff; Potentialfelder der Elektrotechnik; Springer Verlag Berlin; 1932
4. IEC 60156 Insulation liquids – Determination of the breakdown voltage at power frequency – Test method, 2018
5. IEC 60071–1 Insulation coordination, Part 1: principles and rules, 2006
6. IEC 60505 Evaluation and qualification of electrical insulation system, 2011
7. IEC 60071–2 Insulation coordination, Part 2: Application guide, 1996
8. IEC 60296 Fluids for electrotechnical applications - Unused mineral insulating oils for transformers and switchgear, 2012
9. H.P. Moser; Transformerboard, Birkhäuser AG Basel, 1979
10. Andreas Küchler, Hochspannungstechnik, Springer Verlag Berlin Heidelberg, 1997
11. V.M. Montsinger „Belastung von Transformatoren nach dem Grade ihrer Erwärmung", etz Heft 23, S.739; 1931
12. www.huntsman.com
13. IEC 60376 Specification of technical grade sulphur hexafluoride (SF6) and complementary gases to be used in its mixtures for use in electrical equipment, 2018
14. IEC 60480 Specifications for the re-use of sulphur hexafluoride (SF6) and its mixtures in electrical equipment, 2019
15. Mosch, W., Hauschild, W.: Hochspannungsisolierungen mit Schwefelhexafluorid. Dr. Alfred Hüthig Verlag Heidelberg Basel (1979)
16. Gernot Finis, Das Verhalten von Silikongel unter hohen elektrischen Feldstärken, Dissertation Universität Kassel, 2005
17. Oliver Belz, Mikrohohlkugelgefülltes Silikongel als Isolierstoff in der Hochspannungstechnik, Dissertation Universität Kassel, 2010
18. Europäisches Patent EP 2281 294 B1, Hochspannungswandler mit flexibler Isolierung, Patentblatt 2012/04
19. Hanno Schaumburg, Einführung in die Werkstoffe der Elektrotechnik, Teubner Verlag Stuttgart, 1993
20. Roth, A.: Hochspannungstechnik. Verlag Julius Springer, Wien (1938)
21. Hanno Schaumburg, Werkstoffe und Bauelemente der Elektrotechnik Band 5 Keramik, Teubner Verlag Stuttgart, 1994
22. IEC 60672–1 Ceramic and glass insulating materials Part 1: Definitions and classification, 1995

© Springer Fachmedien Wiesbaden GmbH, part of Springer Nature 2022
R. Minkner and J. Schmid, *The Technology of Instrument Transformers*,
https://doi.org/10.1007/978-3-658-34863-2

23. NGK Insulators Ltd., Technical Guide, catalog no. 91, Tokyo, Japan 1974
24. IEC 60672–3 Ceramic and glass insulating materials Part 3: Specifications for individual materials, 1997
25. K. Kunde, J. Schmid, Development of SF6 current transformers and capacitive voltage transformers for high voltage levels with regards to design, installation and service life of instrument transformers, CEPSI (Conference of the Electric Power Supply Industry), Mumbay/India Nov. 2006
26. CIGRE technical Brochure 455, Aspects for the Application of Composite Insulators to High Voltage (≥72kV) Apparatus, CIGRE April 2011
27. Europäisches Patent EP0821373A1, Kunststoffverbundisolator mit spiralförmigem Schirm und Verfahren zu seiner Herstellung, Patentblatt 1998/05
28. IEC 61462 Composite hollow insulators - Pressurized and unpressurized insulators for use in electrical equipment with rated voltage greater than 1 000 V - Definitions, test methods, acceptance criteria and design recommendations, 2007
29. IEC 62217 Polymeric HV insulators for indoor and outdoor use - General definitions, test methods and acceptance criteria, 2012
30. IEC 60137 Insulated bushings for alternating voltages above 1000 V, 2017
31. L. Kuipers und R. Timman; Handbuch der Mathematik, Walter de Cruyter & Co Berlin, 1968
32. H.J. Voss, Technologie der Öl- und SF6-isolierten Wandler, FKH Fachtagung 2003 Hochspannungsmesswandler, Zürich 2003
33. IEC 61869–1 Instrument transformers Part 1: general requirements, 2007
34. IEC 61869–2 Instrument transformers Part 2: additional requirements for current transformers, 2012
35. IEC 61869–6 Instrument transformers Part 6: additional general requirements for low power instrument transformers, 2016
36. IEC 61869–8 Instrument transformers Part 8: additional requirements for electronic current transformers, draft
37. IEC 61869–10 Instrument transformers Part 10: additional requirements for low power passive current transformers, 2017
38. IEC 61869–9 Instrument transformers Part 9: digital interface for instrument transformers, 2016
39. IEC 61850–9–2 Communication networks and systems in substations Part 9–2: Specific communication service mapping (SCSM) – sampled values over ISO/IEC 8802–3
40. R. Boll, Weichmagnetische Werkstoffe; Vakuumschmelze GmbH Hanau; 1990
41. Europäisches Patent EP1851 566 B1, Ringkern-Stromwandler mit Phasenkompensationsschaltung, Patentblatt 2009/27
42. CIGRE technical brochure 57: The paper-oil insulated measurement transformer; 1990.
43. IEC 60060–1 High-voltage test techniques - Part 1: General definitions and test requirements, 2010
44. IEC 60085 Electrical insulation - Thermal evaluation and designation, 2007
45. Europäisches Patent EP0990160B1, Ringkern-Stromwandler mit integriertem Messshunt, Patentblatt 2001/48
46. Ruthard Minkner, Universelle Ringkern-Stromwandler für Mess- und Schutzzwecke, ETZ Elektrotechnik und Automation, VDE Verlag Berlin, Heft 22/2003
47. Ruthard Minkner, Der Drahtwiderstand als Bauelement für die Hochspannungstechnik und Rechentechnik, Vieweg Verlag, Messtechnik, Heft 4 1969
48. Ruthard Minkner, Die Technologie des modernen Hochspannungswandlers, ETG Sponsortagung 22.01.1992 in Muttenz/Schweiz, ETG im SEV Zürich, ETG Band 22d, 1992
49. LOPO – Low Power instrument transformer for medium voltage switchgear, Trench Switzerland Brochure E910, Basel 2011

50. P.A. Nicati, Capteur de courant à fibre optique base sur un interféromètre de Sagnac, PhD Thesis no. 976, EPFL, Lausanne, 1991
51. Klaus Bohnert, Peter Guggenbach, A revolution in high dc current measurement, ABB Review 1/2005
52. Optical current transformers for air insulated substations, Trench Catalog TOCT_IEC, 8/2016
53. IEC 61869–3 Instrument transformers Part 3: additional requirements for inductive voltage transformers, 2011
54. IEC 61869–5 Instrument transformers Part 5: additional requirements for current transformers, 2011
55. IEC 61869–11 Instrument transformers Part 11: additional requirements for low-power passive voltage transformers, 2017
56. IEC 61869–7 Instrument transformers Part7: additional requirements for electronic voltage transformers, draft
57. IEC 61869–15 Instrument transformers Part 15: additional requirements for voltage transformers for DC applications, 2018
58. CIGRE technical brochure 394: State of the art of instrument transformer, October 2009
59. R. Minkner und andere, Ferroresonance oscillations in substations, VDE Verlag Berlin Offenbach 2019
60. Steward, J.L.: Circuit theory and design. Wiley, New York/USA (1956)
61. Liebscher, F.: Wolfgang Held. Springer Verlag Berlin Heidelberg, Kondensatoren (1968)
62. H. Adler., R. Minkner, G. Reinhold, J. Seitz; Ein hochstabilisierter 1.5 MeV Elektronenbeschleuniger; Proc. Eur. Reg.Conf. on Electronic Microscopy, Bd. 1, S. 122 – 125; Delft; 1960
63. R. Minkner; Untersuchungen an Hochspannungsgleichrichtern zur Erzeugung konstanter Gleichspannungen; Dissertation TU Berlin, 1965
64. R. Minkner, E. Schweizer; Low power voltage and current transducer for protecting and measuring medium and high voltage systems; WPRC; Spokane Washington/USA, 1999
65. Menten, L., Amsinck, R.: Electronic Voltage Measurement System for Gas-Insulated Substations; 4th International Symposium on High Voltage Engineering. Greece, Athens (September 1983)
66. P. Matthiessen, U. Weigel; Spannungswandler für Hochspannungsanlagen mit kapazitiven Teilern und elektronischem Messverstärker; Bulletin SEV, Band 71 (1980), S.450 - 455
67. R. Blind; Emil Haefely AG, interner Bericht ZEB Nr. 36 (1972
68. Minkner, R., Schmid, J.: Widerstandstechnologie für Mittelspannungssensoren; ETZ Elektrotechnik + Automation. Heft 17, 18–21 (2000)
69. Chr. Jost, Die Messung der Beanspruchung eines Spannungssensors, Semesterarbeit E5c, University of Applied Sience, Burgdorf/Schweiz
70. Datenblatt: Du Pont 110X High voltage thick film resistor composition, www.dupont.com
71. Ruthard Minkner, Ein universeller RC-Spannungswandler für Hochspannungs-Versorgungsnetze, ETZ Elektrotechnik + Automation; 2002 Heft 22
72. H. Seljeseth, E.A. Saethre, T. Ohnstad, I. Lien, Voltage transformer frequency response. Measuring harmonics in Norwegian 300 kV and 132 kV Power Systems, Harmonics And Quality of Power, 1998. Proceedings. 8th International Conference on Volume 2, Issue 14–18 Oct 1998
73. Ruthard Minkner, Joachim Schmid, Koaxiale Spannungssensoren in gasisolierten Mittelspannungsanlagen, ETZ Elektrotechnik + Automation; 1999 Heft 19
74. M. Pohl, H. Schröder, Einfluss des Ausgleichvorgangs auf den Richtungsentscheid von Distanzschutzgeräten, ETZ-A 90 (1969), Heft 4
75. D. Hou, J. Roberts, Capacitive Voltage Transformer: transient overreach concerns and solutions for distance-relaying, WPRC Spokane, Washington/USA, 1995

76. G. Bnmouyal, D. Hou, J. Roberts, D. Tziovams, The effect of conventional instrument transformer transient on numerical relay instruments, WPRC, Spokane Washington/USA, 2000
77. DIN EN 50160, Merkmale der Spannung in öffentlichen Elektrizitätsversorgungsnetzen, 2011
78. Karl Küpfmüller, Einführung in die theoretische Elektrotechnik, Springer Verlag Berlin 2008
79. Gustav Doetsch, Handbuch der La Place Transformation, Springer Verlag 1956
80. Europäisches Patent EP098003B1, RC-Spannungsteiler, Patentblatt 2006/10
81. Farnoosh Rahmatian, Optical Instrument Transformers, pacworld, Houston/USA, März 2018
82. K. Bohnert, P. Gabus, H. Brändle, Fiber-Optic Current and Voltage Sensors for High-Voltage Substations, 16th International Conference on Optical Fiber Sensors, October 13–17, 2003, Nara Japan
83. Knut Sjövall, ABB Outdoor instrument transformer Application guide, ABB Ludvika/ Schweden 2005
84. IEC 60599 Mineral oil-filled electrical equipment in service - Guidance on the interpretation of dissolved and free gases analysis, 2015
85. User manual IOSK, Trench Switzerland, Basel 2010
86. Omicron MPD 600, www.omicronenergy.com
87. Haefely DDX 9121b, www.haefely.com
88. Praehauser, T.: Elektrotechnik und Maschinenbau. Zeitschrift des Elektrotechnischen Vereines in Wien **86**, S193-201 (1969)

Printed in the United States
by Baker & Taylor Publisher Services